the SEARCH
for TRUTH

the SEARCH *for* TRUTH

Starring

Moses, Darwin & Einstein

John E. Walls

Scripture quotations from the Authorized (King James) Version. Rights in the Authorized Version in the United Kingdom are vested in the Crown. Reproduced by permission of the Crown's patentee, Cambridge University Press.

Scripture quotations from the New King James Version®. Copyright © 1982 by Thomas Nelson. Used by permission. All rights reserved.

Scripture quotations from the Holy Bible, New International Version® NIV®. Copyright © 1973, 1978, 1984, 2011 by Biblica, Inc.™ Used by permission. All rights reserved worldwide.

Scripture quotations from Revised Standard Version of the Bible, copyright © 1946, 1958 and 1971 National Churches of Christ in the United States of America. Used by permission. All rights reserved.

Scripture from the New Revised Standard Version Bible, copyright © 1989 National Council of the Churches of Christ in the United States of America. Used by permission. All rights reserved.

Unless noted otherwise, scripture quotations in the following are from the New King James Version.

Image of Paul Gauguin's painting licensed from Getty Images.

SECOND EDITION

Copyright © 2023 John E. Walls
All rights reserved.

ISBN 978-1-7373654-6-4

Acknowledgements

Thank you Christy for your unflagging support during the years of research and composition that went into this volume. No man ever had a more wonderful daughter. More than watermelon!

I thank J. Garth Hill for writing the *Foreward*. His candid reflections clearly demonstrate the need for this book. Garth is a devout Christian and a Master Mason.

I thank Will Cohen for his review which appears on the back cover of the print editions. I appreciate having his Jewish perspective.

I thank J. Fred Hart, Jr. for his pointed criticisms. Fred is an attorney. I asked him to bring his legal expertise to the task, and he did not disappoint.

I'm indebted to the late Paul Gauguin for his wonderful painting in which he proffered his probing questions: Where do we come from? What are we? Where are we going?

Finally, I thank the scholars for their wonderful works that informed my efforts, especially in Part II. For the most part, I only know them from reading their publications and watching their internet presentations. There were many, but two were especially helpful: Nobel laureate Savante Pääblo, Director of the Max Planck Institute of Evolutionary Anthropology in Leipzig, Germany; and Professor Ian Tattersall, Curator Emeritus of the Museum of Natural History, New York City, New York.

My gratitude to all the scholars in no way implies that any of them would agree with my thesis, my arguments or my conclusions. Perhaps those who are Christian would. Those who are atheist probably would not.

Foreword

When I was in college, I studied both, theology and biology. I thought about ministry and about teaching, but chose another route in the medical field because I was unable to reconcile what I knew about science with what I was learning about theology and the Bible. I tried, but in 1969 the tools were not available to achieve this reconciliation. When I first met John Walls and learned about this book he was writing, I was absolutely excited because after 50 years, here was my possible reconciliation.

After reading the first edition of *The Search For Truth Starring Moses, Darwin and Einstein*, I knew that I had found the lost piece to my faith puzzle that had eluded me so long. I highly recommend this book to anyone studying theology or science who is having trouble with pulling the two disciplines together. This may not be the final answer, but it does allow the reader to finally pose the questions that will lead to the answer, or answers.

When John asked me to write a Foreward to this book, I was greatly honored to be able to do so, primarily because of the help the book gave me. I would recommend that anyone interested in the subjects of religion and biological science take a hard look at this book. It fills in many of the blanks that had been left by one side or the other concerning creationism, evolution or theistic evolution.

Finally, here is a framework on which to build knowledge without denying either, one's faith or, one's scientific learning. This book should be a must-read for first year theology students.

Thank you, John for relighting that candle in the darkness.

— J. Garth Hill

Table of Contents

Prolegomenon	8
Prologue	10
Preface	11
Introduction	13
PART I	**18**
Preamble: Part I	19
Shoulders Of Giants	21
Sitz Im Leben	26
What Is Science	30
Naming	38
Made Of Star Stuff	41
The Big Bang	44
PART II	**54**
Preamble: Part II	55
Evidence Of Others	56
DNA	59
Out Relatives	74
PART III	**80**
Preamble: Part III	81
Dogma	82
The Holy Bible	88
Genesis	94

Truth Claims In Genesis	106
Concerning The Soul	133
Anthropic Principle	143
Theophany & Inspiration	146
The Traveling Toladah	155
Life On Other Planets	168
The Garden In Eden	172
Reproductive Choices	187
The Reckoning	195
The Tree Of Life	202
Life On The Outside	211
Bad Things Happen	223
A Family Tree	226
Gravettian	229
Shoulders Of Giants	233
The Decision	240
Grace	243
Faith	252
Epilogue	260

Prolegomenon

Science without religion is lame.

Religion without science is blind.

— Albert Einstein

Einstein pointed out that science and religion inform each other. If we force them to stand apart we can expect "lame" science and "blind" religion. Given their profound power in our lives, it behooves practitioners to do everything possible to ensure that our science and our religion are first class in every way. One way to do this is to demonstrate the connections between them.

There are many different fields of science and many different religions. For our purposes, the sciences of interest are commonly known as the "Big Bang" and as "Darwinian Evolution." The religion of primary interest is "Christianity."

Every person has an appointment with the Grim Reaper, and when that day arrives, the only thing that will matter is whether their soul goes to heaven or to hell. What could be more important?

Jesus put the issue this way. *"For what profit is it to a man if he gains the whole world, and loses his soul?"*[1] Nothing is more important than our soul landing safe on that happy shore — absolutely nothing!

In the few years we have on this planet we may do many excellent things. We may graduate with high honors. We may win an Olympic Gold. We might marry the person of our dreams and have splendid children. We may become very wealthy. We may even win the Nobel prize. If we accomplish all these things and more, but do not go to heaven when we die, then in the final analysis we are to be pitied.

Drawing upon scripture and science and the connections between them, this book shows the reader how he or she can be certain their soul will arrive in heaven when their body dies. It comes down to a personal decision. This decision is explained in detail near the end of this book.

Prologue

Truth does not need people to understand it, but people need truth in order to really live.

— Professor Peter Plichta

Peter Plichta holds a Ph.D. in physics, a second Ph.D. in chemistry and third Ph.D. in biology. He's earned the right to be heard. People really do need truth because life is orchestrated by decisions, and decisions based on faulty information cannot be expected to produce desirable results. Whether a question is asked in a classroom, boardroom or hospital room, we expect truthful answers. Some truthful answers turn out to be comforting and others disturbing, but regardless of emotions, not only do we want to know the truth, we need to know.

So, what do we mean when we say something is true? Doing so is to make a *truth claim*. A truth claim is an assertion believed to describe the facts of the matter under consideration. History has shown that some truth claims actually are true, but others are false. In order to distinguish between the two, we turn to science and religion, for without looking to these two pillars of civilization, we have little chance of understanding the universe or our place within it. Truth is the treasure we seek, and if a doubt arises about what is — and what is not truth — we adjudicate the matter based solely on the evidence.

Whether the reader is a religious person who would like to better understand the biblical account of creation in light of modern science, or a secular person who rejects the biblical account of creation because of science, the only requirements for a proper reading of this book are an open mind and a firm commitment to follow the evidence wherever it leads.

Preface

> We are all, each in our own way, seekers of the truth, and we each long for an answer to why we are here.
>
> — Professor Brian Greene

Professor Greene is a theoretical physicist at Columbia University and the author of influential books and scientific articles. His aphorism speaks to issues we address in this book.

As seekers of truth, we turn to science and scripture, for we need them both. Great rivers of truth flow from the fountain of science, but as powerful as science is, it cannot answer all our questions.

Each of the three Parts of this book was written to answer one of the three questions French artist Paul Gauguin painted onto his magnum opus. His questions are translated: Where do we come from? What are we? Where are we going? His questions are time dependent. The first two speak to the past. Wherever we came from, we had no say in any of this history. Likewise, whatever we are, we had no say in any of this history either. His third question speaks to the future, and the answer to this question depends on a decision each person will inevitably make, whether they do so intentionally or not.

In Part I of this book, we address Gauguin's first question by examining the truth claims associated with the theory of the Big Bang. In Part II, we address his second question by examining the truth claims associated with the theory of Evolution. In Part III, we answer his third question by comparing and contrasting the truth claims of scripture with the truth claims of science discussed in Parts I and II.

The truth claims in Genesis are central to our task, and because Christians and Jews embrace Genesis as part of their sacred literature, it is fair to say this book is written primarily for members of these two religious communities. However, atheists, agnostics and people of other religious traditions will also find this book useful, provided they bring an open mind and commitment to follow the evidence to the reading. Regardless of one's present religious convictions (or lack thereof) anyone willing to ignore the evidence might as well close this book now and do something else.

In the following, we compare and contrast the truth claims in the first few chapters of Genesis with the corresponding truth claims of modern science. I hope to convince the reader that the comparisons are straightforward, the evidence is unambiguous and the conclusions are clear. Although the vocabulary of science is quite different from that of scripture, the two sets of claims are complementary, *not* contradictory.

The science is presented in a manner meant to be accessible to laymen, which means it is written as simply as possible and without the technical terms associated with various branches of science. Although it has been suggested that scripture and science present different kinds of truth which do not necessarily overlap, the evidence shows there is not one kind of truth from science and a different kind of truth from scripture. The truth claims of science are in harmony with the truth claims of Genesis, and we need them both in order to provide robust answers to Gauguin's three questions.

With the books of science in one hand and the books of scripture in the other, we show that the truth claims of science and the truth claims in the Book of Genesis are synonymous. The either/or approach to science and scripture is no longer tenable.

Introduction

> The truth of our faith becomes a matter of ridicule among the infidels if any Catholic, not gifted with the necessary scientific learning, presents as dogma what scientific scrutiny show to be false.
>
> — St. Thomas Aquinas (1225-1274)

Aquinas wrote long before the Protestant Reformation, so when he used the phrase "any Catholic" he meant any Christian at that time. In his view, an infidel was one who did not believe in God or one who adhered to a religion other than Catholicism.

In the time of Aquinas, dogma was any religious assertion understood to be absolute and unquestionable. This understanding has not changed. Today, his aphorism is applicable for all Christians, including Catholics and non-Catholics alike.

The Book of Genesis makes truth claims about the origin of the universe and truth claims about how the universe has evolved into the marvels we see around us today. Genesis also makes truth claims about the various kinds of life that have arisen on Earth. The Book of Genesis presents its truth claims in outline, and each of the outlines is highly compressed. So, many of the details one might like to see are not included. However, stunning developments in science have provided a host of details about these things. We are now in position to compare what we have learned from science with what we read in the Book of Genesis.

Science can be divided into two general categories: the physical sciences, which are concerned with inanimate objects; and the life sciences, which are concerned with living organisms. In Part I we focus on the physical sciences, and in Part II on the life sciences. As stated above, no special expertise is needed in order to understand the truth claims of science presented in Parts I and II.

Some readers will be more interested in the life sciences than the physical sciences and vice versa. Regardless of the degree of interest the reader has in one or the other of these sciences, he or she is encouraged to stay engaged, and here's why: Without some basic understanding of the sciences presented in Part I and Part II, the comparisons between scripture and science presented in Part III cannot be fully appreciated. Therefore, the information discussed in the first two Parts is prerequisite to understand the comparisons set forth in Part III.

The broad range of topics covered limits the depth of our discussions since a thorough treatment of each topic would result in not one, but a series of volumes. Although much of this book is about science, it is not a textbook of science, per se. Specialists in the various fields of science will find the material in the different sections to be representative of current thought, but far from exhaustive.

For example, one widely accepted truth claim of science is that the human body is made of tiny objects called atoms, and we have unequivocal evidence that there are, in fact, atoms of iron in our blood and atoms of calcium in our bones. In Part I, we turn to the truth claims that arise from the theory of the Big Bang in order to explain the origin of the atoms that make up the human body.

Another widely accepted truth claim from science is that modern human beings have evolved from other, more primitive forms of life. We are now certain that different kinds of bipedal creatures lived on Earth before modern Homo sapiens came along. We know this to be the case, not only because we have some of their fossilized bones, but also because genetic research has clearly shown that modern people carry some DNA from Neanderthals within their own. Therefore, in Part II we turn to sciences such as genetics, paleoanthropology, archaeology and evolution to partly explain how we have come to be the only creatures who read books.

Genesis made truth claims about how the universe has come to be as we find it long before science came along to fill in many of the details. Failure to directly connect the truth claims that arise from a literal reading of scripture with widely accepted truth claims from science leave those interested in these matters in a difficult situation. Unwilling to abandon scripture entirely, some have taken the position that a literal reading of scripture is no longer tenable. This view may have seemed the only alternative in the past, but recent developments in science tell a different story. Even so, there are a number of truth claims from scripture that have no scientific corollary. In such cases, comparisons are not possible. However, in those cases where science and scripture do make a truth claim on the same topic, the two sets of claims are synonymous.

The Genesis narratives describe a number of events that took place in the Fertile Crescent sometime during the last 100,000 years. These events of were not ordinary; they could not have been predicted from, or explained by, what came before them. They were what scientists call *emergent*. These emergent events propelled some archaic hominids to the status of fully modern Homo sapiens. Happily, the science associated with the theory of evolution has provided much of the information we need in order to understand a number of passages that have been confusing in the past. As we read the scriptures in the bright light of science, we see the Genesis account of creation closely mirrors what we have learned from scientific disciplines. We are searching for truth, and the Fertile Crescent is a fertile venue.

Although truth is important for its own sake, not all truth is terribly interesting. Whether or not a particular truth is interesting depends mostly upon the individual. Like beauty, it's in the eye of the beholder. Some truths are riveting to most observers, but other truths are of little interest to anyone. For instance, a truthful answer does exist to the question of which shoe a child put on first this morning, but this particular truth is of no interest to anyone, not even the shoe merchant.

Some questions hold almost universal interest. French artist Paul Gauguin posed three very interesting questions, and his questions form the superstructure for this three-part book.

Paul Gauguin (1848-1903) was born two weeks before the bloody June Days of France's 1848 revolution. His parents decided to leave France after Louis Napoleon was elected president of the Second Republic in December of 1848. The following August, the Gauguin family embarked for Lima, Peru, but during the voyage Paul's father, Clovis, died from a ruptured aneurysm. In 1854 the fatherless family moved back to France to live at Orleans. It was here that Gauguin received a seminary education that influenced the rest of his life, including his art.[2]

At the age of forty-three, Monsieur Gauguin left his wife and children in order to devote the remainder of his life to his art. Prior to his move to the South Pacific, Gauguin shared accommodations with Vincent Van Gogh for a number of months. An argument between the two may have played some role in Van Gogh cutting off part of his own ear. Gauguin moved about for a time, but he eventually took up residence on Tahiti, where, in 1897 he resolved to create a masterwork as his last testament.[3]

As this is being written, Gauguin's masterpiece is on display in the Boston Museum of Fine Art. It's a large painting, measuring approximately five feet high and 13 feet wide.

Each of the three scenes corresponds to one of the questions Gauguin painted onto the upper left corner of his canvas: D'Où venons-nous? Que sommes-nous? Où allons-nous? These are translated: Where do we come from? What are we? Where are we going? The scene on the right depicts three young women, and near them a small baby lying on a blanket: Where do we come from? The center scene depicts a group of adults engaged in various activities against a background of mystical symbols and figures: What are we? The scene on the left depicts an old woman nearing the end of her life: Where are we going?

As tumultuous as his life seems to have been, his questions still speak to us today. Eugéne Henri Paul Gauguin died in Tahiti. He was 55.

PART I

Where Do We Come From?

Preamble: Part I

> Everything that now exists was, in one form or another, contained within the initial fireball that was the cosmos.
>
> — Professor Allen Guth, MIT

Professor Guth is a physicist at Massachusetts Institute of Technology. In his book, *The Inflationary Universe,* he pointed out something quite astonishing. Everything that exists today was, in one form or another, contained within this tiny package.[4] Every material object that exists today, from a very small thing such as an atom of hydrogen, to a very large one such as a galaxy of stars, is either eternal, or it came into existence at some time in the past. If an object actually is eternal, then it has always existed forever into the past, and it will continue to exist forever into the future. If the object is not eternal, then it had a beginning; call it a birthday, if you please. The same observation holds for the cosmos as a whole. It is either eternal, or it has a birthday.

For much of recorded history, many of those who thought about such matters assumed the cosmos was eternal and it consisted only of our solar system, a view that later expanded to include what we call the Milky Way Galaxy. Today we know that both these views were incorrect. We have learned that the universe did have a beginning, and we are certain that the universe is much, much larger than the Milky Way. Our galaxy of some hundred billion stars, even as large as it is, is only a tiny fraction of the whole. There are some hundred billion other galaxies in the visible cosmos, each one containing billions of stars. We have learned the universe first appeared about 13.8 billion years ago as an extremely small, excruciatingly hot and exquisitely dense bundle of energy. From the beginning, the universe has continued to expand, and as it has grown very much larger, it has become very much cooler. The scientific story of this amazing series of events that started about 13.8 billion years go, and continues today, is known as the Big Bang.

The Big Bang has produced a wide variety of objects, and has done so in very large numbers. Early in the expansion, the cosmos was particularly fond of producing hydrogen and helium. These two gases make up about 99 percent of the mass of the observable cosmos today. It also produced a smidgen of lithium. Over millions of years, the primal hydrogen and helium formed the first generations of stars, and over their life times, the stars produced dozens of other kinds of atoms— lots and lots of atoms. There are some ten trillion atoms in a single human body. So, in order to answer Gauguin's first question, we look back in time to the beginning of the universe, for it is here that the atoms which are now reading this sentence got their start. However, before we explain how we know what we think we know, we briefly recall some of the milestones laid down by those who came before us.

Some readers are not very interested in the history of science. Such as these may want to skim over or entirely skip the next chapter. This history is not required in order to understand the comparisons which follow. However, understanding (as best we can) the science of the Big Bang and of Evolution is necessary in order to appreciate the comparisons provided in the subsequent chapters.

Shoulders Of Giants

If I have seen further than others, it is by standing upon the shoulders of giants.

— Sir Isaac Newton

We are very fortunate that we do not have to start on 'page one' of the book of science with each new generation. As we pursue our scientific investigations, we really do stand on the shoulders of giants. In my opinion, Sir Isaac Newton was the greatest scientist who ever lived. Many others have made remarkable contributions to science and written eloquent volumes explaining their insights, but Newton's contributions were unprecedented. As did Newton, we also stand on the shoulders of giants when we look out over the landscape of science.

Human beings want to understand the events that shape their lives, and in ages past they believed that floods, droughts, hurricanes, earthquakes, disease, volcanoes and other disasters were supernatural events. So, ancient people invented various gods to explain the things they could not otherwise understand. In these ancient mythologies, if the gods were pleased, things went well. If the gods were displeased, less pleasant things happened.

This way of thinking about nature began to change around the time of Thales of Miletus (circa 624 BCE - circa 547 BCE). He laid the foundation for the idea that nature functions according to certain principles. Thales' insights began to overthrow the idea that an oligarchy of capricious gods caused earthly disasters.

A bit later Parmenides of Elea (circa 500 BCE) wrote, "Being is un-generable and imperishable." His idea that matter does not unpredictably appear or disappear laid the foundation for what are known as the *conservation laws*.

Democritus (circa 460—370 BCE), another Greek, believed that the cosmos contains indivisible components from which everything else is made. He reasoned that if one cuts a piece of copper into smaller and smaller pieces, eventually no further cutting would be possible. He concluded that material objects were composed of very small invisible particles with nothing but empty space between them. He named these particles, *atoms* from the Greek, *atomos* (*a* — not) and (*tomos* — cuttable). If one were forced to leave a single chapter from the book of science to some future, but scientifically ignorant civilization, it would be difficult to find anything more crucial to the development of their science than atomism. Democritus is called the father of atomism.

We suppose most people have heard of the Greek philosopher, Aristotle. His influence was dramatic, and unfortunately, he disagreed with Democritus. Aristotle taught that everything is made from some combination of earth, air, fire and water, immersed in an all pervasive aether. His towering influence was such that his view was taken as superior to the view of Democritus, even though, as we now know, Aristotle's view was in error. Unfortunately, Democritus' budding flower of atomism could not bloom in the deep shade cast by Aristotle. Thus, the ideas of Democritus were placed on hold for centuries. One lesson to be learned from this account is that no one is immune from error. The views of even highly respected individuals such as Aristotle are sometimes just wrong.

In the first century BCE the notion of continuity in existence was given a more robust form by Titus Lucretius Carus in his 6 volume poem, *De Rerum Natura* (On the Nature of Things). Lucretius, as he's usually called, wrote, "Nothing can be created from nothing." Professor Lawrence M. Krauss is a distinguished physicist at the Large Hadron Collider. In his 2012 book Professor Krauss wrote, "… not only can nothing become something, it is required to do so."[5] Lucretius believed, "Material objects are of two kinds, atoms and compounds of atoms. The atoms themselves cannot be swamped by any force, for they are preserved indefinitely by their absolute solidity."

The atomic theory was revived in the work of the English scientist, John Dalton, who in 1803 produced the first table of atomic weights. Investigators began to notice a certain periodicity in the elements that were known at the time, and Russian chemist, Dmitri Mendeleev (1834-1907) developed a chart of rows and columns in which he grouped those with similar properties in logical sequences. This 'Periodic Table of Elements' is usually on display in chemistry classrooms across America.

A number of years later, Nobel laureate, Joseph John Thomson (1856 — 1940) discovered negatively charged particles which he named *electrons*. Using his observations of electron behavior, Thomson developed a model of the atom sometimes likened to a plum pudding. In this model the individual plums were analogous to his newly discovered electrons, and the rest of the atom was the pudding with a positive electrical charge evenly spread throughout. In this manner the atom was electrically neutral. Another analogy for the Thomson model could be that of a loaf of raisin bread in which the raisins represent the electrons and the surrounding loaf represents the positively charged cloud around the electrons.

The plum pudding model of Thomson was shown to be incorrect through the work of Sir Ernest Rutherford who directed *alpha particles* (positively charged helium nuclei) at a thin sheet of gold. He discovered that instead of passing through in a random fashion as would be expected in the J. J. Thomson model, some alpha particles were deflected back to the detectors which he had positioned around the experiment. These results were taken as conclusive proof that very small positively charged areas were present within the gold foil. Rutherford had detected what we call the *nucleus* of the atom. The nucleus is the tiny, dense, positively charged center of the atom. Electrons orbit about the nucleus. Rutherford's model was a good improvement over the plum pudding model, but it still was incomplete

According to what was known about classical electromagnetic theory at the time, the negatively charged electrons moving in orbit about the positively charged nucleus would radiate energy in a continuous spectrum, and as they did so, they would quickly spiral down into the nucleus. This was not happening, so investigators knew that the Rutherford model needed refinement.

Neils Bohr was a student of Thomson, and Bohr solved the issue of orbiting electrons decaying. He proposed that the electrons could only occupy certain stable orbits, each corresponding to a precise energy level of the electron. This refinement served to prevent electrons from spiraling ever downward into the nucleus, something everyone concerned knew did not occur. In the Bohr model of the atom, electrons could stay in some stable orbit indefinitely, but they also could jump from one orbit to another by emitting or absorbing specific amounts of energy. In other words, when a particle of light (a *photon*) of sufficient energy is absorbed, the electron can jump to a higher level, and a photon of the same energy is given off when the electron drops back into the less energetic orbit. Bohr's model matched the evidence available at that time. Nevertheless, more refinement was necessary in order for the model to explain the evidence that was accumulating.

In the early twentieth century, scientists discovered that the nucleus of almost all atoms contain more than only protons. They also contain particles called *neutrons*. So, at this point in the story, the atom consisted of protons and neutrons in the nucleus along with electrons in orbit. Evidence continued to accumulate that suggested yet another level of complexity must exist within the nucleus.

Murray Gell-Mann and George Zweig (independently) proposed that nucleons (protons and neutrons) contained even more elementary particles. Zweig named the new particles aces. Gell-Mann called them *quarks*. Gell-Mann received the Nobel Prize for his new quark model, and his choice of the name for the new particles is the one that stuck.

The above sketch contains only a highly abbreviated outline of the development of the atomic theory. There are many other particles such as *muons* and *mesons*, as well as *antiparticles*. The current list of particles also includes the mysterious *neutrino* which permeate the cosmos. It may make you feel a little strange if you think very long about it, but some trillion neutrinos pass through your body every day. Let it be enough for now to acknowledge that it has taken a long time and the shoulders of many giants to elevate us to our present level of understanding. So, we tip our hat to those who provided us a more lofty view.

One final observation: even though we are blessed to see further by standing on the shoulders of giants, we must be careful that our giants are facing the right direction.

We are about ready to begin comparing and contrasting the truth claims of science with the truth claims of scripture, but before we do, we need to familiarize ourselves with something called *Sitz im Leben*.

Sitz Im Leben

> With the passage of time, every human being accumulates a set of assumptions, prejudices and biases that inevitably influence how he or she understands the world.
>
> — Professor Lee R. Berger

Professor Berger is a well known anthropologist, and his observation just above will suffice to introduce *Sitz im Leben*. This German term is difficult to translate precisely into English, but *situation in life* is a fair approximation. It is very important to understand the power of Sitz im Leben, for it inevitably influences how we understand different ideas that arise, not only from science, but also from scripture.

Sitz im Leben was coined by Hermann Gunkel. The term was employed by Julian Wellhausen around the turn of the nineteenth century in connection with the belief that the ancient biblical texts could not be properly understood without understanding the communities that produced them. These scholars came to believe the first five books of the Old Testament, known as the Pentateuch, are a compilation of other documents. This idea became known as the *documentary hypothesis*. Wellhausen believe he could identify these imaginary documents based largely on the name of God each one employed. Julias Wellhausen is sometimes called the father of the documentary hypothesis. Hermann Gunkel cautioned, "… theologians should learn that Genesis is not to be understood without the aid of the proper methods for the study of legends."[6]

For our purposes, we need not delve any deeper into the particulars of the documentary hypothesis at this point. However, we very much need to understand the power of Sitz im Leben, for new ideas can threaten those who have built their careers expounding the orthodox view. Those who have busied themselves deepening the ruts established by those who came before them may feel their reputation, or even their livelihood, might be jeopardized by some new idea. When it comes to evaluating new ideas, the fly in the ointment is Sitz im Leben.

For our purposes, Sitz im Leben will denote the particular set of assumptions, prejudices and biases held by an individual, rather than the more general set shared by a community. Sitz im Leben is more insidious than some would expect and certainly more powerful than we might like to believe. The growth of Sitz im Leben is cumulative. Rather like the gradual buildup of barnacles on the hull of a ship, these assumptions, prejudices and biases are barely noticed at first, but if left unattended, they will eventually become problematic. Barnacles can be removed from a ship, and poor assumptions, prejudices and biases can be removed from our Sitz im Leben, provided we realize they are present, and decide they must go. Perhaps the discovery of a new piece of evidence may cause an individual to understand things differently. A famous example of the power of the Sitz im Leben can be seen in the events that took place during the development of the theory of the Big Bang.

Albert Einstein's equations of general relativity are the framework for our present understanding of how the cosmos came to be as we find it. But in their original form, Einstein disliked what his equations revealed. They showed that the universe had a beginning, and Einstein's Sitz im Leben included the assumption that the cosmos must be eternal. As brilliant as Einstein was, and in spite of the fact that he, above all others, should have taken his original equations literally and seriously, he did not, at least, not at first. In order to force his equations to agree with his Sitz im Leben, he did something extraordinary; he added an arbitrary term to the original equations. He called this term the *cosmological constant*. It was a fudge factor, and its only purpose was to make his equations agree with his notion that the universe did not have a beginning. Such was the power of his Sitz im Leben in action. The 20th century's most famous physicist was not immune to the influence of his Sitz im Leben, and neither are the rest of us, immune to the influence of ours.

In the spirit of fairness and in order to clarify this episode, we point out that as the evidence continued to accumulate, Einstein scrubbed this particular barnacle from his intellectual ship, if you will. He later called his addition of the cosmological constant his "greatest blunder." Due to the influence of his or her Sitz im Leben, anyone can unwittingly accept a spurious interpretation of science or of scripture, but as did Einstein, the great ones follow the evidence, even when they don't much like where it leads.

Progress in science will be hindered if we fail to take its truth claims literally and seriously. In his superb book, *The First Three Minutes*, Nobel laureate Stephen Weinberg, after explaining the missed opportunities in the early days of the theory of the Big Bang, pointed out that such opportunities might have been seized upon if theorists and experimentalists had been in closer communication and if theorists had taken their own work more seriously. He wrote, "This is often the way it is in physics - our mistake is not that we take our theories too seriously, but that we do not take them seriously enough."[7]

Indeed, our search for truth will be hampered if we fail to take seriously what our theories tell us, and the same observation applies to how we approach scripture. However, if we do take the creation stories of Genesis literally and seriously, and then compare and contrast them with modern science, we can develop answers to Paul Gauguin's three questions. These answers will be based solely on evidence.

Weinberg's book is a real treasure. His Chapter 11 should be mandatory reading for all seminary students, not for the conclusions, but for the astute analysis of the state of religion in these times. His analysis accurately describes problems addressed in this book. When taken literally and seriously, science and scripture form a treasure trove of answers to our deepest questions.

Numerous other examples of the influence of Sitz im Leben can be found in many areas of life, but we are concerned in Part I and Part II with its influence on our understanding of science, and in Part III, to an even greater extent, on our understanding of scripture. Spurious interpretations of the first few chapters of Genesis are riddled with assumptions, biases and prejudices. Some of these interpretations are not supported by serious and literal reading of scripture, and they are robustly refuted by modern science. We'll say more about the influence of Sitz im Leben on interpretation of scripture in Part III.

Science is one of Homo sapiens' finest accomplishments. It permeates our lives, and we are very much indebted to those who have taught us so much about the cosmos and ourselves. A catchphrase heard in these days of the COVID-19 pandemic is "follow the science." Since we have such unbridled confidence in science, it behooves us to say a few words about what science is, and perhaps even more importantly, what science is not.

What Is Science

> Science is far from the objective and impartial search for incontrovertible truths that nonscientists might imagine. It is, in fact, a social endeavor where dominating personalities and disciples of often defunct yet influential scholars determine what is 'common knowledge.'
>
> — Nobel laureate Savante Pääblo

Those who doubt laureate Pääblo's insight only need to look back to Aristotle's condemnation of Democritus' theory of atomism to settle the matter. So, how might we explain this social endeavor we call science? Well, explaining science can be a bit slippery, not only because of the broad range of subjects that scientists investigate, but also because of the different techniques they employ in their pursuits.

Perhaps a metaphor will help. We can imagine science as a grand tapestry woven from threads of many colors. As we look at this beautiful wall hanging, we notice bright red threads of astronomy, forest green threads of geology, silvery threads of chemistry and golden threads of genetics. We also notice other colored threads representing biology, evolution and anthropology, and so on.

At the center of our tapestry is a sunburst of royal purple threads radiating outward in all directions. The sunburst represents the science we call *physics*, and it is royal purple because physics is the king of all science. Lord Rutherford reportedly said, "All science is physics; the rest is stamp collecting." If this is an accurate quote, his words probably offended a number of those engaged in other fields of science.

Feelings aside, physics is the king of science, but it is also the servant of the other branches of science, as the numerous intersections of the threads in the tapestry suggest. Each thread in our tapestry is different from all the others. But together, they glisten and glimmer as they reflect humanity's progress in the search for truth.

Although the different threads share some common characteristics, those useful in one field may be of little value in another. On the other hand, some threads that are primarily associated with one field of science may prove very useful in others. For example, discoveries in anthropology are rarely (if ever) cited in papers on nuclear physics, but dating techniques derived from physics have proven invaluable to anthropologists in their efforts to develop timelines of human evolution. So, scientists weave various threads together to form the tapestry of science, and although the threads are quite different in some respects, in other respects they function to reinforce each other. But, our metaphor can take us only so far.

There are different ways of doing science, and obviously, there are many products that derive from science. Given this diversity, the question of what science actually is can be simplified by focusing on two of three themes, each one a bit different from the others, but not too far distant. It's rather like three peas in a pod, if you will. The first pea in the pod is *purpose*; the second pea is *process* and the third pea is *product*. We will focus on the purpose and the process of science, since the products of science are abundant and need no further explanation.

Nobel laureate Eric R. Kandel has written, "The aim of science is to discover new truths about the world..."[8] His is a concise answer to the question of the purpose of science. He uses the word, aim, but if we understand him correctly, the meaning is the same as purpose. In his view, the purpose of science is to discover new truths about the world. Kandel makes the following assumptions.

First, he assumes truth exists. Some students, when first introduced to what scientists call the *uncertainty principle,* have been heard arguing that objective truth cannot exist, and if they are correct, then scientists are a sad lot because we search in vain for that which cannot possibly be found. However, such students are mistaken because they fail to understand the limits of the domain within which the uncertainty principle is applicable.

Second, Kandel assumes that through science we can come to know some measure of these new truths, and the historical record of science leaves no room for doubt on this point.

Third, his phrase, "to discover" implies that there are at least some truths that are presently unknown, and we are certain that this is the case. After all, if we already knew everything (or believed we did), why would we continue to search?

Defining the process of science should be approached with a measure of caution because one field of science may use a process that is of little value in another field. For instance, the processes associated with nuclear physics are very different from those associated with anthropology. In addition, we can be cautious in our approach because certain assumptions about the processes are misguided. For example, the uninitiated may assume that every step in the scientific enterprise is totally objective, and although this is an admirable goal, it turns out that this assumption is not valid. It is evident, or at least it should be, that each new investigation begins with a subjective choice of what to investigate. In addition, the process is, at least at first, guided by preconceived ideas and influenced by bias toward explanation.[9] These are troubling and unavoidable consequences of being human, but even more troubling is the issue of honesty.

Nobel laureate Richard Feynman pointed out that although people usually believe they are honest, in fact, they are not. He said that scientists are not honest in as much as honesty requires that one make clear the entire situation, to provide all the information required for somebody else who is intelligent to make up their.mind.[10] Failure to disclose all the relevant information will, in the final analysis, be problematic in any field of inquiry. In spite of our limitations, objectivity is certainly a lofty goal in science, and we proceed on the basis that scientists are honest and objective in their work. In the spirit of Feynman, we must point out that most of the "stuff" in the cosmos is beyond our current understanding.

A well known aphorism states: seeing is believing. And it turns out that in order to interpret any piece of scientific evidence, we really do need to see something. It may be light from a distant star through a telescope, or light from a small object through a microscope, or perhaps data on a computer screen. It may be that what we see is produced through special detectors such as those at the Large Hadron Collider in Europe. Whatever form the evidence takes, we need to see it in order to believe it. The particulars of how we see the evidence is not the issue. But some sort of experience on the part of the observer is essential to the scientific enterprise.

The problem is there is a whole lot of "stuff" in the cosmos that we have no way of seeing, at least not yet. We call this invisible stuff *dark matter* and *dark energy* precisely because we cannot see it. We can train our telescopes on the area where we know the stuff is, but we simply cannot see it. We know the dark matter and the dark energy are real because we can see the effects they produce on other things that we can see.

The curious reader may wonder if the dark part and the observable part of the cosmos are comparable in size. Not even close! The dark part of the cosmos makes up about 95 percent of the total, while the other 5 percent is observable. It's the 5 percent of the cosmos that we have good reason to believe we know something about. Currently, we have no idea what the dark part actually is, but we do believe that it is approximately 70 percent dark energy and 25 percent dark matter. This limitation on our knowledge can be compared to a banker who can account for only one nickel of each dollar on deposit, or to a stock broker forced to admit that she has no idea where 95 cents of every dollar is invested. Dark revelations, indeed. One might assume that our ignorance of the dark part of the cosmos would serve very well as a firewall against hubris, but it turns out that this lack of understanding provides little protection against speculative ideas that cannot be tested against evidence. For instance, there is not one scrap of evidence to support the idea that ours is only one among many other universes. As concerns this issue, we really are in the dark.

In his 2001 book, *Our Cosmic Habitat*, Professor Martin Rees writes the following, "Dark matter is the No. 1 problem in astronomy today, and it ranks high as a physics problem, too. If we could solve it — and I'm optimistic that we will within the next decade — we would know what our universe is mostly made of..."[11] Optimism offers a cheery outlook, and it is preferable to pessimism, but here we are, *two* decades later, and we know the dark stuff is real. But we still don't know what it is. (An optimist might claim the glass is half full. A pessimist might claim the glass is half empty. A theologian might claim it's the wrong glass. A careful physicist will claim the glass is completely full: half liquid and half gas.) The point is that most of the content of the cosmos is hidden from view, and the following information about the cosmos is limited to the 5 percent we can see.

So, the scientific process begins with a subjective choice of what to investigate, and it continues with speculation and conjecture as the scientist imagines a plausible solution she thinks may explain the phenomenon under consideration. If the scientist believes her solution has merit, she may formulate a more rigorous explanation known as a *hypothesis*. In order to qualify as authentic science, the hypothesis must make predictions that can be tested against evidence. In this manner, the hypothesis is subject to falsification. Along the way, the scientist may construct some sort of experiment intended to garner evidence that will either support or disprove her hypothesis. If the predictions are supported by the evidence, then the hypothesis may be promoted to the next level of confidence known as a *theory*.

A scientific theory is more than only an educated guess. It is a logically consistent explanation that makes predictions. The predictions make the theory falsifiable. So, when we speak of the theory of Relativity, or the theory of the Big Bang, or the theory of Evolution, we are not simply guessing.

Does this mean that scientific theories are beyond doubt? No, it does not. If only one piece of evidence comes to light that contradicts the theory, then the theory must be improved or discarded completely. Every scientific theory remains subject to falsification by the evidence, and history has shown that even old and respected theories must at times be improved or abandoned in light of new evidence.

Improving or culling incorrect theories is an integral and important part of the scientific process. However, if some theory has been successfully tested against all the available evidence, and if the scientific community accepts the theory as common knowledge, then it may be elevated to the next level of confidence known as a scientific *law*.

The laws of nature are believed to be universal, which means they operate throughout the cosmos in the same manner they operate on earth. Scientists group these laws into categories such as the laws of thermodynamics, the laws of motion and so on. One such law that has proven invaluable for the advance of science is the law of *conservation of energy*, which we can state very simply in the following manner: The energy content of the cosmos is constant. Even though energy can be converted from one form into another, it cannot be created or destroyed.

We convert the chemical energy of gasoline into mechanical energy that moves our automobiles. We convert electrical energy into light for our homes and businesses, and so on. We have the ability to manipulate different forms of energy, but we cannot increase or decrease the cosmic total.

Albert Einstein taught us that mass and energy are equivalent, and as mentioned above, Alan Guth of the Massachusetts Institute of Technology (MIT) pointed out that everything in the cosmos today was, in one form or another, part of the Big Bang. This means everything we see today (including the Earth) was part of the Big Bang. So, the energy in the cosmos today is equal to the energy that was present in the beginning.

It strains our minds to imagine that the energy of trillions of stars was contained within something smaller than the head of a pin as the Big Bang claims, but this theory is substantiated by evidence. Science depends on the laws of nature, for they describe a regularity that can be understood, and thankfully so. If the operation of the cosmos was capricious, we would have little chance to develop evidence based theories which explain the cosmos. We understand some of the laws which regulate the cosmos, but we know that we do not know them all. We have learned the laws of nature are real, and they operate across the universe without our permission.

It bears repeating that each scientific episode begins with a subjective choice of what to investigate and continues with imagination as the scientist formulates a hypothesis. The hypothesis is little more than a plausible model which seems to explain the phenomenon under consideration. The hypothesis must make predictions that can be tested against evidence, and if the predictions are substantiated by the evidence, then we may call it a scientific theory. A theory whose predictions have passed all known tests against the evidence may be elevated to a higher status known as a scientific law. Science is one of Homo sapiens' finest accomplishments; no other species has been capable of doing science.

Even though the early stages of investigation include imagination, speculation and conjecture, these three are not science, per se. Boundaries between what is and what is not science are to be maintained. According to one prominent scientist, "No one should blur the distinction between what is well established and what is conjectural."[12] We've said a good deal about science, not only because science is so very important, but also because, as Nobel laureate Savante Pääblo pointed out, science is not what some imagine it to be.

To tidy up just a bit, one may believe her latest and greatest hypothesis explains some phenomenon, but if it makes no predictions that can be tested against the evidence, then the hypothesis, no matter how beautiful or how elegant, does not qualify as a truly scientific hypothesis. Predictions that are testable against the evidence are a non-negotiable part of science, and evidence is the final judge of any scientific hypothesis. When the hypothesis is tested against the evidence, if the evidence and the hypothesis do not agree, then regardless of its apparent beauty, and regardless of the author's credentials, the hypothesis is just wrong.

Naming

"What's in a name?"

— William Shakespeare

Shakespeare penned the question, and as the story unfolds, the families of Romeo Capulet and Juliet Montague despise each other — a fictional conflict reminiscent of the real feud between the Hatfield and McCoy families in American history. Young Romeo and Juliet are in love and determined to follow their hearts, regardless of family objections. In the famous "balcony scene" Juliet says to Romeo, "What's in a name? That which we call a rose by any other name would smell as sweet." The tragic death of the young lovers leads to the resolution of the conflict between their families. So, do names really matter? Well, names mattered a great deal to the families of Romeo and Juliet.

People today come from a long line of hominids who lived in small tribes without family names. These ancestors of ours didn't know how to name anything for millions of years, but they understood their dire need to be an accepted member of their small tribe. Food, water, shelter, protection and breeding opportunities were directly related to one's tribal status. Today, we name all sorts of things, including the tribe we identify by a family name.

Vestiges of our tribal ancestry can be seen in modern society. Members of a particular tribe tend to defend their members against those of different tribes. Hatfields defended their tribe, and McCoys defended theirs. This tribal mentality is still alive. We call some tribes Democrats, some Republicans, some Independents, some Catholic and some Protestants. We call some Communists, some Muslim and others Jewish. Depending on one's tribal affiliation, other tribes may be accepted or vilified.

One negative way this tribalism manifests itself is in various criminal organizations and gangs that threaten those not in their tribe. There are hundreds of such tribes. Negative examples include: MS-13, Bloods, Crips, Mafia, Yakuza and Cosa Nostra. Positive examples include: Free Masons, Lions Club, Kiwanis Club, Catholics, Methodists, Baptists, and so on. Tribal names matter, and so do given names. One rarely meets a man named Judas or a woman named Jezebel.

For the last few thousand years, fully modern Homo sapiens have been assigning names to all sorts of things. Whether we understand a thing quite well or barely at all, we'll name it. In general, it is customary to allow the person who discovers something new to name it. Not only do we assign names to physical objects, but also to abstract ideas and emotions.

We have the ability to divide the world into abstract symbols (names) which describe how things are. We can also recombine the names in different ways to imagine how things might be different. The ability to communicate names and ideas through complex language and written symbols make us different from other kinds of life. No other species has these capabilities. Exactly how naming works is less clear than the fact that we do it.

In Part III, we'll say more about the importance of our capacity to name things. For the present, as Aristotle suggested would be helpful, we will define a few important terms (italicized) before we continue. As mentioned above, Aristotle was wrong when he hindered the development of atomism, but he got a lot of other stuff right.

First, the words *cosmos* and *universe* are synonymous. They each denote not only the gigantic container we call *space*, but also the enormous number of objects within the container. Obviously, everything we observe is in the universe we inhabit. We have no evidence for anything that is not part of this universe. The idea that there may be other universes, possibly an infinite number of others, is pure speculation. But let that go for now.

Second, the words *observable* and *dark* denote two distinct categories of things that are part of this cosmos. As mentioned earlier, the part of the cosmos we can see directly, we call observable; the part which we cannot see directly, we call dark. For instance, with the unaided eye, we can see this sentence. We can see the moon and Sun and a limited number of stars in the sky. With the aid of various kinds of telescopes, we can see distant galaxies. However, the cosmos also contains things we cannot see, not even with all our advanced technology. Although we cannot directly observe the dark part of the cosmos, we know it is real because of the gravitational effect it has on other things that we can see. Still, we cannot directly observe the dark matter itself — that's why we call it dark. Unless noted otherwise, when we use the words cosmos or universe we will be referring to the observable part. So, the cosmos contains two distinct parts: the observable and the dark.

Third, material objects such as people, planets and stars are made from tiny bits of matter called *atoms*. Atoms are made from three particles we have named *protons*, *neutrons* and *electrons*. The electrons are thought to be *fundamental* particles, which means they cannot be divided into smaller particles. However, the protons and neutrons are *composite* particles, which means they are a combination of other more basic (fundamental) particles. These other fundamental particles which combine to form protons and neutrons are called *quarks*. We'll say more about atoms just below, but for now, let it be enough to realize that the material objects of the observable cosmos, including the human body, are all made from very small bits of matter that we call atoms.

Fourth, the body is made of small biological containers we call *cells*. There are about 200 different kinds of cells in the body, and almost all of them have a small core called the *nucleus*. We'll go into more detail about cells in Part II. With these few definitions, let's continue.

Made Of Star Stuff

The nitrogen in our DNA, the calcium in our teeth, the iron in our blood, the carbon in our apple pies were made in the interior of collapsing stars. We are made of star stuff.

— Carl Sagan

The stuff Carl Sagan mentioned were atoms. Atoms are tiny conglomerates of protons, neutrons and electrons. The protons and electrons have equal and opposite electric charges. Protons have a positive charge, and electrons have a negative charge. Neutrons don't have electric charge. The protons and neutrons are called *nucleons* because they bind together inside the tiny, dense, central area of the atom we call the *nucleus*. Electrons orbit about the nucleus.

Nature has given us 92 different kinds of atoms. The number of protons in the nucleus make each atom whatever kind it is. Nitrogen has 7 protons, carbon has 12, oxygen has 16, calcium has 20 and iron has 26. Atoms other than the simplest form of hydrogen also have some number of neutrons in the nucleus.

Carl Sagan was correct, and it is widely known that the human body is made of atoms. But then, so are stars, planets and apple pies. Unequivocal evidence shows that from oak trees to honey bees, every material object in the observable cosmos is made from atoms.

High school students are usually taught that there are atoms of iron in their blood and atoms of calcium in their bones. However, not every student leaves high school with a thorough knowledge of the kinds of atoms that make up their own body, and even though a robust accounting of the atoms is more interesting, it is less well known.

Beginning with the most abundant atom in the human body, here's the full account: We are approximately 65% oxygen, 18% carbon, 10% hydrogen, 3% nitrogen, 1.5% calcium, 1% phosphorus, 0.35% potassium, 0.25% sulfur, 0.15% sodium, 0.05% magnesium, with trace amounts of copper, zinc, selenium, molybdenum, fluorine, chlorine, iodine, manganese, cobalt, iron, lithium, strontium, aluminum, silicon, lead, vanadium, arsenic and bromine.[13] Some of these atoms are highly poisonous, but in trace amounts, they are either necessary for life as we know it, or at least not particularly harmful.

So, there are 28 different kinds of atoms in the human body, and lots of them. About 7 billion, billion, billion atoms come together to form a person who tips the scales at 70 kilograms (about 154 pounds). These atoms group together in various ways to form some ten trillion cells. It may be disquieting to learn that the body also contains comparable numbers of bacteria.[14] Bones are made of atoms, and so are bacteria. Where did these different kinds of atoms come from?

We have discovered that every material object that exists today, from a very small thing such as an atom of hydrogen, to a very large one such as a galaxy of stars, is either eternal, or it came into existence at some time in the past. This principle holds for the cosmos as a whole; it is either eternal, or it has a birthday. This observation invites us to look more closely into the Big Bang, a theory that describes the sudden appearance of the cosmos about 13.8 billion years ago and the cosmic expansion that continues today. The theory explains how the cosmos produced atoms of different kinds in such huge quantities, but the theory tells us nothing about how, over billions of years, various groups of these atoms came together to form many kinds of living things, including those who walk about on two legs and ask questions.

It is unclear whether Albert Einstein or Lord Rutherford made a comment about being able to explain science to a barmaid. It is possible that neither one said anything along this line. Whether the comment is apocryphal or historical, the idea is that no special training should be necessary to understand the laws of physics if they are presented in a simple manner. The following outline is meant to do just that, and if we are successful, then the layperson will understand where the atoms in her own body came from.

The cosmos is not eternal, and neither are any of the things inside. Remember, Part I speaks only to the question of where the atoms of the human body came from. Other topics, such as evolution and soul, will be taken up in Parts II and III.

Some readers may not be very interested in the science of the Big Bang, but the material in Part I is essential when it comes to comparing the opening chapters of Genesis with what we have learned from science.

The Big Bang

"The evolution of the world can be compared to a display of fireworks that has just ended: some few red wisps, ashes and smoke. Standing on a cooled cinder, we see the slow fading of the suns, and we try to recall the vanishing brilliance of the origin of the worlds."

— Reverend Georges LeMaître

As is obvious from his title, Reverend Georges LeMaître (1894-1966) was a clergyman. What is not obvious is that he was also a brilliant mathematician who earned a Ph.D. from the Massachusetts Institute of Technology. Unlike some of his contemporaries, he was familiar with the very important advances in astronomy taking place in his time.

Albert Einstein's equations of general relativity provided the framework for our present understanding of how the cosmos came to be as we find it, and Reverend LeMaître was among the first of those to solve Einstein's original equations, no small feat in itself. More importantly, LeMaitre was the first to accept and properly interpret the implications of the solutions. This clergyman taught us that the universe suddenly appeared in what he poetically likened to a display of fireworks.

Nothing in the universe is eternal. This is true for cats and cockroaches, as well as for planets and people. It's also true for the universe as a whole. Name whatever object you will, this principle holds. For centuries there were some, including Einstein and Fred Hoyle, who assumed the cosmos was eternal. But today we have a solid theory, supported by very good evidence, that tells a different story. We call this story the Big Bang. We are quite certain the cosmos did have a beginning, and this is arguably the most important scientific discovery ever made. Those who believed the universe to be eternal were mistaken.

The labors of many scientists have confirmed that the universe came into existence about 13.8 billion years ago, and that it has been expanding ever since. We are forever indebted to Albert Einstein for his equations of Relativity, and to Reverend Georges LeMaître for his solution and proper interpretation of the equations in their original form.

We have learned that near the beginning, the cosmos was much smaller than the head of a pin, very much hotter than the interior of any star and packed more tightly together (more dense) than seems possible. Under these extreme conditions, our physics breaks down, which means we cannot accurately describe what things were like at the first instant. However, our physics works pretty well just a few billionths of a second after the beginning.

So, what was happening shortly after the beginning? Well, $E=mc^2$ is the only equation in this book, and it's probably the most famous one in the world. With this simple equation Einstein taught us that energy (E) and mass (m) are equivalent, which means that under certain conditions, either one can convert into the other. The temperature near the beginning of the Big Bang is estimated to have been about 100 million, million, million, million, million degrees.

Even though we cannot accurately describe what things were like at the very beginning, we believe that very soon after the first moment, the cosmos was filled with fundamental particles called *quarks* and *leptons*, and with energy. Since quarks and leptons are fundamental, they cannot be broken into other components. In light of Einstein's equation of mass/energy equivalency, my conjecture is the quarks and leptons distilled out of the unimaginable energy that suddenly came forth out of nothing. Quarks and leptons are the building blocks natures uses to assemble more complex materials such as the atoms of calcium in our bones and the iron in our blood. From the bottle of Scotch on the shelf behind the bar to the bartender who pours the Scotch into a glass on the bar, it's all made from quarks and leptons.

The early cosmos also contained enormous numbers of other fundamental particles known as *photons*. These are particles of light, and they outnumbered the other particles by about 1,000,000,000 to 1. In the young, hot, dense cosmos, the other particles were bathed in astonishing light.

Temperature is a measure of energy; things that are hot wiggle, jiggle and fly about more quickly than things that are cold. The higher the temperature, the faster the particles move. For example, molecules of boiling water are moving more quickly than molecules of cool water. This principle applies to particles in the early cosmos. Due to the astonishing temperature in the early stages of the cosmos, everything was wildly flying about. This soup of very energetic particles we have been describing is known as *plasma*.

It turns out that the photons of light interact strongly with particles in the plasma. This means light can't pass through plasma very well at all. In other words, plasma is opaque to photons.

The cosmic maelstrom was so violent that the fundamental particles could not combine to form more complex particles. However, the cosmos was expanding, and as it continued to expand, it continued to cool, which means the particles were slowing down. When they had slowed enough, different kinds of quarks combined to form *protons* and *neutrons*. These two kinds of particles form the dense, central region of the atom called the *nucleus*. Due to the positive electric charge of protons, the nuclei of atoms have a net positive charge.

Those not very interested in science might begin to daydream about now, but hang on. Understanding this science is necessary in order to appreciate the comparisons of science with scripture in Part III. Be encouraged; the effort will be worth it. With regard to plasma and photons, perhaps an analogy will help.

Light passes easily through a glass window, but suppose a window is replaced with a mirror. In this case the light will be reflected back instead of passing through. Electrically charged particles swarming about in the early cosmos acted something like gazillions of tiny mirrors. When the light interacted with these charged particles, it was reflected in random directions; scientists say the light was *scattered*. The photons were not immobilized. They were just sent off in some other direction, over and over again. The charged particles were so numerous and so dense, that the light couldn't travel very far before it scattered off in a different direction. Although the cosmos was filled with astonishing light, it was dark because the light couldn't shine.

Another kind of lepton called an *electron* was part of the young cosmos. Electrons have a negative electric charge equal in magnitude to the positive charge of protons. About 380,000 years after the beginning, the cosmos had cooled enough for electrons to enter orbits around the nuclei. This was a major milestone in the evolution of the cosmos. Up to this point, the photons of light had been trapped inside the plasma of charged particles, so the cosmos was dark. However, when the electrons entered their orbits, the negative charge of the electrons balanced the positive charge of the nuclei. In this manner the cosmos became electrically neutral and transparent to the light.

At this moment in the expansion, the light was released from the expanding hypersphere of plasma. We have discovered this ancient light fills the cosmos today. We call it the *cosmic microwave background*, CMB for short. The CMB is strong evidence that the theory of the Big Bang is accurate. The cosmos, which had been dark for about 380,000 thousand years suddenly became transparent to the astonishing light within.

Unless you happen to be a scientist you probably don't think very much about light, except, perhaps, when you see a rainbow or when you think those unpleasant thoughts at night when the driver in the oncoming car won't dim his headlights. We may also think about light when we put on makeup or when we take photographs, but understanding light can be a little confusing.

In the ordinary sense, when we use the word light, we refer to the light which enables us to see. But the light we can see with our eyes is a small segment of a much larger spectrum of electromagnetic radiation, most of which we cannot see. Mobile phones send out and receive electromagnetic radiation when in use, but, - and thankfully so - we cannot see this kind or radiation. Unless we are situated in some very isolated location, cell phones and their users are constantly bathed in the electromagnetic radiation coming from phones and the towers they use. The light we can see with our eyes is a small part of a much larger spectrum of electromagnetic radiation, and all of it has some peculiar properties.

For one thing, electromagnetic radiation in a vacuum travels 186,000 miles each second. It's faster than anything else we know. Light is so fast that it could travel about seven and a half times around the Earth in one second. Nevertheless, light is not infinitely fast, which means it takes time for light to get from one place to another. As the distance to be traveled becomes larger, the time required for the trip increases. For instance, it takes about eight minutes for light to travel the roughly 93,000,000 miles from the Sun to the Earth. The universe is so big that it takes billions of years for light to travel to Earth from some far away galaxies. So, when we examine light from very distant objects, we are seeing into the past.

Light carries information as it speeds along, and very often, we can examine the light and determine where it came from and what generated it. In addition, we have learned how to code information into the electromagnetic radiation (the light) on the one end, and to decode the information on the other. This is how cell phones, televisions, radars and radios work.

Light travels easily through empty space and through the window panes in our homes, but the light we see with our eyes does not pass through our clothes. However, X-rays are electromagnetic radiation more powerful than the light we can see with our eyes, and X-rays can pass through our bodies. This type of light enables physicians to take images of things inside our bodies. Still, the light we can see with our eyes cannot pass through even a very thin sheet of aluminum foil.

For another thing, light behaves differently at times. Sometimes it behaves like a wave, and at other times, like particles. This strange attribute is sometimes referred to as the *dual nature* of light. Albert Einstein not only taught us that mass and energy are equivalent, he also taught us that all kinds of light (electromagnetic radiation) comes in packets of energy called *quanta*. For a visual, we can imagine these quanta as something like pearls strung on a tightly stretched string and all moving at 186,000 miles per second in the direction the string is pointing.

Our string of light travels in a straight line; this is why we can't see around corners. The closer together the individual pearls are on the string, the more energetic the light. For example, blue light is more energetic than red. This means the pearls of blue light would be more closely spaced along the string than the pearls of red light. X-rays are more energetic than both, so the pearls (quanta) of the X-rays would be more closely spaced than the blue or the red, and so on. With our eyes, we can see some of these quanta, these strings of pearls, but not others.

To tidy up just a bit: we have learned that the early cosmos produced enormous quantities of electrically charged particles, and the photons of light outnumbered the other particles by about a billion-to-one. As the cosmos expanded and cooled, the various particles combined in different ways to produce atoms of hydrogen, helium and a tiny bit of lithium. The young cosmos produced so much hydrogen and helium that these two gases account for about 99 percent of the mass in the observable universe today. Gargantuan clouds of these gases are found in the Milky

Way Galaxy and throughout the universe. However, the Big Bang did not directly produce the heavier atoms.

As discussed earlier, the heavier atoms were created inside stars. Here's an overview of this process. When a gas expands it cools down. When it is compressed, it gets hotter. (This is the principle behind modern air conditioning.) Over millions of years following the beginning, the force of gravity compressed segments of the giant clouds of hydrogen and helium gases, and as gravity squeezed the gases into smaller and smaller volumes, the gases grew hotter and hotter. The compressed gases eventually got hot enough to support a process known as *nuclear fusion*.

Fusion occurs when protons and neutrons fuse together to form heavier particles. When particles do fuse, a tiny amount of energy is released. Stars are huge. The number of fusion events inside stars is just staggering, which means that the total energy output from a star is enormous. In the nuclear maelstrom within its core, the Sun fuses six hundred million tons of hydrogen into helium every second. Through nuclear fusion in its core, the Sun produces about 5,600 million million million million calories of energy every minute! At this rate of consumption, some might wonder if the Sun is going to run out of fuel and leave Earth to freeze. Not to worry; the Sun is so huge that it will take a few billion years more to use up its hydrogen, but eventually, it will. When it does, the Sun will begin to fuse helium into heavier atoms. This process continues in stages, and every stage produces a different kind of heavier atom such as carbon, oxygen and iron. So, the heavier atoms such as carbon, oxygen and calcium found in the human body were created inside stars. Okay, but how did these heavier atoms get from the stars into the human body?

Well, stars come in different sizes, and some of the larger ones end their lives in a massive explosion call a *super nova*. When such stars explode, they blast the heavier atoms they had already produced back into space where they seed other clouds of hydrogen and helium. Over hundreds of millions of years more, the force of gravity compresses portions of these enriched clouds into second generation stars such as our Sun. Some of the enriched material involved in this process of star formation can remain outside the newly forming star, and this enriched (dusty) material can form planets, moons and other things, all orbiting around the star. Not only do we live on one of these dusty old planets, our bodies are made from the dust. So, the atoms that make up the human body were created inside stars, then blasted into space where they eventually formed other stars and planets. The atoms of the human body really do come from the dust of the Earth. As Govert Schilling said, "We are made from the stuff of stars that was blown into space billions of years ago by energetic stellar explosions, and we are an inexorable part of the cosmic cycle."[15]

Even so, it takes more than just a collection of atoms to form a human being. It takes the proper environment. We must have food to eat and water to drink, and so on. It turns out that the Earth is just the right distance from the Sun to have favorable temperatures. The Earth receives the proper amount of energy for green plants to grow and to keep the planet warm enough to have liquid water. If we were much closer to the Sun, it would be too hot; much further away and it would be too cold. The Earth is located in what is sometimes called the 'Goldilocks Zone' - not too hot and not too cold. We have discovered that the atoms of the human body were created inside stars, and one star, the Sun, supplies just the right amount of energy to sustain life here on Earth. How lucky is all this?

Lucky or not, a lot of things absolutely must come together in just the right way in order for human beings to exist. The cosmos is what it is, and it does the things it does because of the different kinds of particles it contains and the forces that act upon them. We have learned a good deal about both. We have measured the mass and the electric charge of many particles. We have measured the force of gravity. We find these values to be constant, and there are dozens of such physical constants. Because they have the values they do, the universe takes on the configuration it does. In his book entitled, *Just Six Numbers,* Professor Martin Rees points out that our universe is very finely tuned, and that changing this tuning, even slightly, would dramatically change the cosmos. He argues that altering any one of the six numbers he discusses would profoundly alter the history of the cosmos and prohibit life as we know it.[16] I'll say more on these issues in Part III, The Anthropic Principle.

The first atoms, the hydrogen and helium and a smidgen of lithium, were created in the early cosmos, and the heavier atoms such as calcium and iron were created inside stars. In this view, the short answer to Gauguin's first question is this: We came from the Big Bang by way of the stars. Even so, it requires more than just a big bunch of atoms to make a human being. The atoms must be processed through a long series of biological contingencies we call *Evolution*.

Now we come to Paul Gauguin's second question: What are we?

PART II

What Are We?

Preamble: Part II

I make no claim of special expertise in Part II. It is written from the perspective of an interested layman drawing from the work of experts in various sciences. Without the contributions of experts in fields such as anthropology, archaeology, biology, evolution, genetics and so on, we would have little chance of formulating a satisfactory answer to Gauguin's second question.

Two things that dramatically changed the way we understand ourselves occurred within a span of only three years in the nineteenth century. Skeletal remains of Neanderthals were discovered in Germany in 1856, and in his 1859 book, *Origin of the Species*, Charles Darwin introduced us to an idea we call the theory of evolution. These two things not only marked the end of the idea that Homo sapiens is the only species of hominid to have ever lived on this planet, but they also gave us the information we need to understand a number of previously enigmatic passages in the Book of Genesis.

Those of us who greatly appreciate science *and* scripture find ourselves in a lovely position. Finally, we have the information we need to objectively compare and contrast the truth claims from scripture with those of science, and having done so, to intentionally make the decision that will determine where our soul goes when we die.

Evidence Of Others

We are committed to following the evidence, and indisputable evidence has shown that several species of hominids arose inside Africa over the last few million years.[17] A partial list of these early Africans includes: Homo erectus, Homo habilis, Homo ergaster, Homo heildelbergensis and Homo sapiens. Homo sapiens show up in the fossil record inside Africa about 300,000 years ago. The evidence is clear; we have some of their fossilized bones.

In addition, researchers from Tel Aviv University have recently identified a new type of early human at the Nesher Ramla site. These fossils have been dated between 140,000 to 120,000 BCE. This new species shares features with Neanderthals and archaic Homo, but it is quite unlike modern humans. The scientists refer to this new species as 'Nesher Ramla Homo type'. They suggest these folks mated with Homo sapiens. They also suggest that the Nesher Ramla Homo type was an ancestor to archaic Homo populations in Asia and to Neanderthals in Europe. In their view, Neanderthals did not evolve in Europe, but in the Eastern Mediterranean area known as the Levant. Professor Rachel Sarig said, "As a crossroads between Africa, Europe and Asia, the Land of Israel served as a melting pot where different human populations mixed with one another, to later spread throughout the Old World. The discovery from the Nesher Ramla site writes a new and fascinating chapter in the story of humankind"[18]

Neanderthals show up in the fossil record outside Africa at around 400,000 years ago. There have been no Neanderthal fossils discovered inside Africa or in the Americas. Neanderthals were tough, brawny folks who, at times, lived under very harsh conditions. Ice ages came and went during their tenure. Mount Toba exploded around 73.5 thousand years ago, and this volcanic explosion may have plunged the Earth into what scientists call a volcanic winter. Furthermore, the Earth is surrounded by a magnetic field which protects us from much of the harmful radiation from the Sun. Recent evidence has shown the magnetic North and South Poles of the Earth swapped places about 41,000 years ago. This reversal event lasted about 1,000 years, and increased radiation reaching the Earth from the Sun during this period may have contributed to Neanderthal extinction. The last Neanderthal died sometime around 30,000 years ago.

We have discovered there were several different kinds of hominids leaving their footprints in the dust over the last few million years. As mentioned above, the earliest of these creatures show up first in Africa. One kind of these African ancestors is called *anatomically modern* Homo sapiens because they had bodies much like modern people. They were not what anthropologists call *fully modern* Homo sapiens because their cognitive abilities were not on par with modern people. From Africa, some of these anatomically modern Homo sapiens immigrated into the Middle East sometime around 100,000 years ago. Genetic evidence shows that during their travels, some of the anatomically modern Homo sapiens from Africa interbred with the Neanderthals.

For the most part, anatomically modern Homo sapiens were creatures of instinct, not symbolic thinkers. Slight improvements in their primitive stone tools may suggest a gradual improvement in cognitive skills, but nothing comparable to the dramatic cognitive leap that took place after they left Africa and before about 23,000 years ago. Although Neanderthals didn't change very much during their long tenure, the evidence shows that anatomically modern Homo sapiens experienced an unprecedented upgrade. Somehow, the early version of Homo sapiens, the ones we call anatomically modern, made a huge cognitive leap that thrust them into fully modern modality. The anatomically modern folks are extinct. We, the fully modern Homo sapiens, are the only species that can recognize, understand and rely on evidence to answer Gauguin's questions. Somehow, the anatomically modern folks became us.

Even though Paul Gauguin posed three questions about one kind of life, he did not ask about life in general. We assume that most people have some idea of what life is, but to develop a formal definition of life is not as simple as one might expect. Laymen may be surprised to learn that not all scientists agree on a single definition of life. Debate continues on what life is and what life is not, and if the experts do not agree, we certainly will not presume to offer a definition of our own. For our purposes it is enough to acknowledge that life exists, and humans are one kind of life among the many others.

We are still unraveling our enigmatic history, and as sketchy as the evidence is for exactly how we reached our present biological pinnacle, no one doubts we are, (to borrow the title from Professor Ian Tattersall's extraordinary book) *Masters of the Planet*.

Part II does not fully answer Gauguin's second question for it contains no information about Adam and Eve, and any answer that ignores the contribution of this couple will inevitably be incomplete. Even so, we will confine our discussion in Part II to the Darwinian evolution of life in general, before proceeding in Part III to explain the contributions of Adam and Eve in particular.

DNA

Nobel laureate Erwin Schrodinger gave a series of lectures in 1943 which became a book the next year.[19] The title of the book is *What is Life?* He was not only concerned with how life seems to violate certain laws of physics, but also with the information that directs life. The great physicist believed there must be something he described as a "non-repetitive, 'code-script' heredity molecule that contains the entire pattern of the individual's future development and of its functioning in the mature state."

Schrodinger was correct; there is such a molecule. We call it *deoxyribonucleic acid* which we abbreviate DNA. As molecules go, DNA is large and hugely complex. It has to be, for it contains all the information necessary to construct a human being and to direct all the chemistry that goes on inside the body. We cannot say for certain, but DNA may be the most complex mechanism on Earth. So, how did the discovery of the heredity molecule Schrodinger predicted in his lectures come about?

As the story is told, Maurice Wilkins, while working at King's College London, obtained a number of images of the DNA molecule between 1948 and 1950. Roslind Franklin came to King's College London in 1951. Raymond Gosling was working on his Ph.D. under the supervision of Franklin in 1952, and Gosling managed to take an image of the DNA molecule. This image, known as Photo 51, was produced by X-ray diffraction.

Pictures taken with X-rays may sound a bit like science fiction to some, but most of us have seen X-ray images of our teeth taken at a dentist's office. The technique for making dental images are not exactly the same as those used to produce Photo 51, but each of these uses X-rays to create images.

Obviously, Gosling made an image of something much smaller than a human tooth. It seems that Wilkins showed Photo 51 to Francis Crick who, along with his colleague James Watson, worked out the structure of the DNA molecule. Wilkins, Crick and

Watson shared the 1962 Nobel Prize for their work. So, we tip our hat to Erwin Schrodinger for predicting this molecule of heredity, to Gosling for photographing it, and to Crick, Watson and Wilkins for describing the structure of this amazing molecule that orchestrates the biological symphony we call life.

Everyone has DNA in their body, and when a man and a woman have a child, the infant inherits half of his or her DNA from the father and half from the mother. The entire package of inherited DNA is called the *genome*, and although most of the information in one person's genome is identical to the information in all others, no two are exactly alike. Slight differences in the genome cause each person to be unique.

I have claimed this book is written simply enough for laymen to understand, but DNA is a very large molecule, and it contains huge amounts of information that regulate thousands upon thousands of processes. To go much further into the operation of DNA requires a number of technical terms. So, in order to fulfill the promise of simplicity, and still provide enough details to satisfy those with some previous exposure to science, we divide the material about DNA into three sections.

Section A is a very brief outline on the structure and function of DNA. Section B provides a few more details. Readers who feel they have enough information from reading Section A, should feel free to skim or skip Section B. Section C speaks to the part of DNA called the epigenome. Some familiarity with how DNA works is required in order to understand why fully modern Homo sapiens is the most dangerous species this planet has ever hosted. We cannot formulate a good answer to Gauguin's second question without knowing a bit about DNA.

DNA: Section A

For a thing (any thing) to exist, it must be somewhere. Where is the DNA molecule located inside the body? Well, about 7 billion, billion, billion atoms come together to form an average size person, and these atoms group together in various ways to form trillions of tiny biological units known as *cells*. Cells that determine gender are called *gametes*. The other cells in the body are called *somatic*. The male gamete is called a *sperm*, and the female gamete is called an *ovum*. When a sperm and ovum come together, a new and unique molecule of DNA is produced. Almost all cells in the human body have a copy of DNA within a small central region called the *nucleus*. This is the same word used to denote the tiny, central part of an atom, but the nucleus of an atom is not to be confused with the nucleus of a cell. These are different entities located in different places.

Some might find it disquieting to learn that the body also contains comparable numbers of bacteria.[20] It is well known that the human digestive tract contains huge numbers of bacteria, some beneficial, but others harmful. There are so many bacteria in the human body that it's rather like a biological war going on inside each person every day. Most of the time we are unaware of this conflict, but at other times, there's no ignoring the fact that some bacteria have made us ill. Some of these harmful bacteria have the potential to kill us.

Bacteria are not the only hitchhikers on the highway of humanity; viruses also catch a ride. (A new virus named COVID-19 has killed millions of people across the earth in the last few years.) Fortunately, the DNA located inside the nucleus of cells also directs the body's biological defense systems, without which the bacteria and viruses would kill us all, and that rather quickly.

DNA can be likened to a cookbook. Chefs understand that recipes in cookbooks call for certain ingredients, meant to be combined in correct amounts, in the proper sequence and cooked at the proper temperature for a specified length of time. If the recipe is followed, the chef might produce a culinary delight. But if she puts a cup of salt into the mixture instead of a teaspoon, or if he cooks the ingredients for 3 hours instead of 30 minutes... well, you get the picture. Sections of the DNA molecule that encode the recipes required to build particular materials are called *genes*. The DNA cookbook contains millions of recipes, and these recipes must be followed precisely or poor results will follow.

Proteins are major components of the body. We get some proteins from the foods we eat, and the body manufactures many others as needed. Without any permission or conscious effort on the part of the owner, when the body needs a certain protein, molecules within the cells read the DNA and set the biological machinery to work to produce what is needed. Obviously, the recipes in an English cookbook are written in words formed from the 26 letters of the alphabet, along with a few numbers. DNA has its own alphabet, and it is very short, consisting of only four letters. Amazingly, DNA uses these four letters to spell out the recipes necessary for building thousands of various proteins the body needs from time to time.

What does DNA look like? Well, it's shaped a bit like a twisted ladder. Ladders have two rigid side-rails with a number of rungs that connect the two. Ordinary ladders are usually made of wood or aluminum or fiberglass. DNA also has two side-rails, sometimes called *backbones*. The DNA backbones are flexible and long enough to accommodate billions of rungs between the two. The two backbones of DNA are made of only two kinds of molecules, and the rungs are made from only four other kinds, which we abbreviate as A, T, C and G. Each rung consists of two of the these four. The As pair with the Ts, and the Cs pair with the Gs. Each of these pairings (the A-T and the C-G) forms one rung between the two backbones. The information in DNA is encoded by the As, Ts, Cs and Gs, not by the individual molecules themselves, but by the sequence in which they occur along one of the backbones. The flexible DNA molecule, with its billions of rungs, is twisted into a spiral, and since there are two backbones, DNA is sometimes referred to as a double helix.

Most cells contain a copy of the DNA molecule inside their nucleus. The exception to the rule is mature red blood cells, which have no nucleus and therefore no DNA. Without DNA they cannot repair or reproduce themselves. Even though the total package of hereditary material is called *the genome*, calling it 'the' genome can be somewhat misleading. There is no single, universal genome, which means there is no such thing as 'the' genome. Although most of each person's genome is identical to every other person's genome, no two genomes are exactly the same. Regardless of how we define life, it is orchestrated by DNA inside the nucleus of most cells.

Not only does DNA control how the body is structured and how it functions, it also plays a role in behavior. It does not hold complete control in all cases. If it did, our actions would all be predetermined by the DNA, in which case there would be no 'free will' and thus, no personal responsibility for our actions. Even so, it does hold virtually absolute control in other cases. For example, individuals whose DNA limits their height to 4'11" will almost certainly never play professional basketball. Individuals whose DNA results in a very low IQ are highly unlikely to be part of a team who designs the next supercomputer or the next Mars Rover.

Environmental considerations also affect behavior; it's the 'nature versus nurture' debate. Both are important, but there are situations within the animal kingdom where the environmental conditions just don't matter very much. For example, say a Shetland pony is kept in the same barn, eats the same foods, drinks the same waters, is groomed with the same brushes and exercised like the thoroughbreds in the other stalls. Even under these conditions, it would be rather foolish to place a bet on the Shetland pony to win the Kentucky Derby.

DNA is the famous code-script molecule which physicist Erwin Schrodinger predicted must exist, and the entire package of information is encoded by the billions of rungs on the DNA ladder. This molecule contains a lot of information, and not all the information is required for any particular task. As mentioned above, the information required for the various tasks are packaged into units called genes. This is why we call the full package of information the human *genome*.

Human beings have an estimated total of 20,000 to 25,000 genes which are packaged into bundles of varying sizes called *chromosomes*. Human beings have 46 chromosomes which combine to form 23 pairs in most cells. *Gametes* (sperm and ovum) contain 23 chromosomes each. When sperm and ovum unite, the two sets of 23 chromosomes manage to form pairs with their counterparts and a new life begins. Chromosomes come in different shapes as well as sizes. The 23rd chromosome is the smallest, and it's the one that determines if a person is male or female. Whether one is tall or short, has blue eyes or brown, has black hair or red or blonde and all the rest, it's all regulated by the genes we inherited from our parents.

So, when a sperm cell unites with an ovum, the DNA from the mother combines with the DNA from the father, and this union produces a unique new life which scientists call a *zygote*. The zygote contains all the information needed to regulate not only how many and what kinds of cells are produced, but also where the cells are positioned in the body. Obviously, brain cells need to end up inside the skull. Two eyes, one nose and one mouth need to be located on the front of the face, and teeth need to be located inside the mouth, and so on. DNA regulates what the cells do, how they grow, how they reproduce, how they repair themselves and everything else. Because each zygote grows into a unique person, DNA evidence is admissible in a court of law.

For example, DNA recovered from a crime scene can establish for certain whether or not a person of interest has been there. Samples of the DNA of the accused are compared with the samples from the crime scene, and if the two of them match, then there is no question; the person of interest has been at the scene of the crime. Like a fingerprint, each person's DNA is unique.

There you have it: A whistle stop tour of how DNA from each parent combines to form a unique molecule of DNA that contains all the information necessary to build a Neanderthal, an Einstein or a barmaid. If this enough for you, dear reader, feel free to skim or skip Section B. It is written for those with a bit more exposure to science, and it contains more technical information.

DNA: Section B

Cells are small globs of a jellylike substance called *cytoplasm* enclosed within a *lipid* membrane. The membrane not only holds the individual cells together and gives them their shape, it also regulates what goes into the cell and what comes out — ingress and egress. The trillions of cells that form the human body come in about 200 different varieties, and almost all have a nucleus. The nucleus is enclosed by its own distinct membrane, and the large molecule of DNA resides inside the nucleus. Cells also contain several other types of structures outside the nucleus, each one enclosed within its own separate membrane. These intercellular bodies are called *organelles,* which means 'small organs.' A cell is a container with different kinds of smaller containers inside.

The DNA directs the construction of different kinds of cells, and these cells group together to form the body's various *organs* such as the heart, brain, liver and so on. The DNA also regulates repair and replacement of cells as needed.

With regard to the nucleus within cells, mature red blood cells are the exception; they have no nucleus and no DNA. Without DNA, the red blood cells cannot repair or reproduce themselves. This is a problem since red blood cells are crucial to life, and they are not very durable. They undergo *apoptosis* (die) after about 121 days. Red blood cells are small in comparison to most other cells, and they account for roughly 25% of the total number of cells in the body. They are so small that a teaspoon of blood contains about

25 billion red blood cells. Their small size enables them to pass through tiny blood vessels called *capillaries* in order to deliver their precious cargo of oxygen to tissues that need it and to carry away carbon dioxide, a waste product.

Since red blood cells cannot repair themselves, replacements must be produced in huge numbers, but where? We have discovered that red blood cells are manufactured in the bone marrow. Bone marrow contains the necessary instructions to make replacements, and it makes them in remarkable quantities. The bones in our bodies release about 2,000,000 new red blood cells every second.[21] Bone marrow is a very busy place, but then, the entire human body is a beehive of activity.

As they go about their business, cells can be damaged. However, all is not lost. We are thankful a broken bone can knit itself back together and a skin laceration can heal. In fact, at least for a few years, most cells can repair and reproduce themselves as necessary to maintain the body in working condition. We are wonderfully made, but eventually our bodies wear out, and we die. Regardless of how many blood cells our bones produce, sooner or later our organs fail, and when they do, the atoms of our body go off to do other things.

Let's consider how the body manages to maintain itself while we are still here. We can say that *proteins* are the main substances the body uses to form structures and to carry out the millions of chemical reactions that take place in the human body every day. There are about 100,000 different proteins the body uses in many, many ways. These proteins are made of *amino acids* strung together rather like tiny links of sausages. It is commonly reported that there are 20 of these *proteinogenic* (protein creating) amino acids, but this is not entirely accurate. There is at least one more amino acid in the human body known as *selenocysteine*. Many more amino acids exist in nature than those manufactured from the DNA code, but the human body gets along pretty well by producing just 20.

The information required to manufacture a protein is encoded in DNA, and when required, the information is copied by a molecule of *ribonucleic acid* (RNA) and transported outside the nucleus to another organelle known as a *ribosome*. The ribosome reads the information from the RNA and then assembles the strings of amino acids in the right order to make the required protein.

Another organelle of interest is called a *mitochondrion,* and there are hundreds of *mitochondria* inside most cells. The mitochondria also contain a molecule of deoxyribonucleic acid which we designate as *mtDNA*. This molecule of DNA is much smaller than the one inside the nucleus. The nuclear DNA ladder contains about 3.2 billion rungs. The mitochondrial DNA ladder contains about 16,500. Another difference between the two is mtDNA is inherited only from the mother, whereas nuclear DNA is a combination of the genes from both parents.

Another organelle is called the *Golgi apparatus*. The Golgi apparatus packages the proteins assembled by the ribosomes for transport to wherever they are needed. The Golgi can be likened to the person at the market who sacks the groceries and delivers them to the customer at home. Cells also contain *lysosomes* and *peroxisomes* which function as recycling agents. These digest foreign bacteria that invade the cell, and they recycle worn-out cell components. This is not a full list of organelles, but it will suffice for our purposes.

Now, back to our ladder analogy to explain a few more details. As stated earlier, the DNA molecule incorporates just six different materials. Two of these, the *phosphates* and the *deoxyribose*, form bonds to create the side rails (backbones) of the ladder, and the other four combine in one of two ways to create the rungs between the backbones. In order to get some visual idea of this backbone structure, let the letter O represent the deoxyribose, and let the letter P represent the phosphate. In this visual, the DNA backbones would look like this: O^P^O^P^O^P^O^P repeating the same alternating pattern, hundreds of millions times. DNA contains two of these backbones, and the two are connected by rungs made of *adenine, thymine, cytosine* and *guanine*, abbreviated as A, T, C and G, respectively. These A, T, C and G molecules are called *bases*. As discussed earlier, these four bases bond with each other in two specific ways. Adenine bonds with thymine to form the A-T pairing, and cytosine bonds with guanine to form the C-G pairing. We can imagine these pairings rather like two friends walking along while holding hands. Each friend will have one hand holding the hand of their companion and their other hand free. In this analogy, the free hand of one of the A-T friends connects to one of the backbones, and the free hand of the other friend connects to the other backbone. When the connection between the two backbones is formed, it creates one rung on the DNA ladder. The same things apply to the C-G pairs. So, each of these paired bases constitute a single rung, and again, there are billions of rungs on the DNA ladder.

The string of alternating O and P molecules that form the DNA backbones carry no information by themselves. Instead, the information is encoded in the bases (A,T, C and G) that form the rungs strung between the backbones. However, this can be a bit confusing at first glance. The information is not written in the rungs, per se. Instead, it is written in the three-letter sequences such as ATA or CGT or ATG, as read along just one of the backbones. These sequences of three letters are called *codons*, and the combination of any one of these letters in a codon with its corresponding O-P segment of the backbone is called a *nucleotide*. The information is hidden until the DNA molecule unzips between the paired bases which form the rungs.

So, the nucleotides read along one of the backbones, encode the information required for the assembly of *amino acids*, and the amino acids, strung together in the proper sequences, form *proteins*. This mechanism for encoding information can be compared with how English words use three letters to convey different meanings. For instance, consider the three letters T, C and

A. If they are arranged in the sequence CAT, they describe a furry feline that some people keep for pets. If the same three letters are arranged ACT, they give an entirely different meaning. Sequences of such triplets which occur along one of the backbones encode the instructions necessary for ribosomes to assemble amino acids into different kinds of proteins, and the body uses these proteins to grow and repair cells as needed.

There is some redundancy built into the code. Although each triplet (codon) codes for one - and only one - amino acid, more than one triplet can code for the same amino acid. For examples, the codons AAA and AAG encode for the amino acid *lysine*, and GCA, GCC, GCG and GCT each encode for the amino acid *alanine*.

Not only is the information in the DNA molecule very precious, it is also a bit fragile. If it gets damaged, bad things can happen. DNA is protected to a large degree by its physical structure. The codons such as the ACG and GCC are enclosed within the double helix structure which is zipped together as the As, Ts, Cs and Gs form their pairs. The ladder-like, twisted helix, then folds over and over onto itself, and portions of the DNA molecule are wrapped around bodies called *histones*. This arrangement is a bit like thread wrapped around a spool or fishing line wound on a reel. The configuration protects the interior of the molecule from most negative influences, but it also hides the information from ordinary usage.

Cells must replicate themselves in order for the body to grow and for cell replacement. In order to do these things, each cell (other than red blood cells) must have a copy of the DNA. The DNA ladder can split itself by unzipping the A-T and C-G pairs. Since the As connect with Ts and Cs connect with Gs, when the DNA molecule unzips between the connected pairs, each of the two halves can reconstruct the other to produce two new complete molecules of DNA. When the cell divides, each of the two daughter cells contain their own molecule of DNA. This is how DNA is replicated when cells are replicated.

Somehow the body knows not only *which* proteins are needed, *when* they are needed and *where* they are needed, but also *how much* is needed to perform a given task. When a piece of information from inside the DNA molecule is needed to make a protein, something has to happen to expose the encoded and protected information, and it does. To use our previous analogy, the section of DNA that holds the required information unwraps from around the histones to open the cookbook to the right chapter. The section of DNA that holds the exact recipe unzips between the A-T and C-G pairs. After it unzips, a molecule of RNA then moves along the exposed segment and copies the information in the recipe. It knows it has reached the end of the recipe and should stop copying when it reaches what is called a *stop codon*, for instance TAA, TAG, and TGA. As discussed earlier, the information is then transported outside the nucleus to the ribosomes. These assemble the amino acids into the required proteins. A bit later, the DNA molecule once again zips up the exposed area until it is needed again, and the cookbook is closed.

Genes are not always available for the body to read. Rather like light switches, genes can be turned on and off as needed. This is called *gene expression*. This amazing DNA molecule somehow knows which genes to turn on in order to make a needed protein and when to turn the gene back off again. Some mechanism is at work to regulate which, when and for how long a particular gene is expressed. All this, and more, takes place automatically as needed. The body does not ask the owner's permission to do any of this. What an amazing molecule!

DNA: Section C

The Epigenome

Large segments of the 3.2 billion nucleotides in DNA do not directly code for proteins, and scientists used to call the non-coding material 'junk' DNA. Frankly, we didn't understand the role of the junk DNA, and we didn't view this portion of DNA to be very important in the grand scheme of things. We thought it was genetic trash, but we were mistaken. Nowadays, that which used to be called junk is called the *epigenome,* and it is hugely more important than first believed. We have discovered that the epigenome orchestrates gene expression with high precision, permitting humans to develop and adapt to their surroundings.[22]

Previously, there was no reason to suspect that experiences could change a person's DNA. However, scientists have discovered that traumatic experiences do change the epigenome, which in turn changes the gene expression in the person involved in the experience. What's more, these epigenetic marks regulate not only growth and development, but also behavior.[23] The most startling part of these discoveries, at least to me, is that these changes do not end with the victim of the trauma. Not only do traumatic experiences change the genome of the person directly involved in the experience, but the changes can be inherited by their progeny.

Concerning trauma, we can barely imagine the circumstances of our early ancestors. They lived with the constant threat of being eaten by animals on four feet, and at times, by those on two. These early hominids almost certainly heard the screams and saw the spectacle of some member of their tribe being caught and eaten. The epigenome of those traumatized in such ways was changed, and this bears repeating: epigenetic changes, sometimes called *marks,* can be passed from parent to child. So, an individual's junk

DNA (their epigenome) is not only conditioned by the traumatic events in their own life, but also by traumatic events experienced by their predecessors. The implications of these discoveries are staggering! In light of this new information, sayings such as, 'He's a chip off the old block' or 'The apple doesn't fall far from the tree' make more sense. When it comes to understanding how the body works, we have discovered the epigenome is definitely not junk, and it has been shaped by trauma over many generations.

As in any other hugely complex process, mistakes within the genetic code do occur from time to time. These mistakes are called *mutations*. Any mutation in the DNA molecule may produce a change that damages the body, but it can also produce a change that some might consider desirable. For example, a genetic mutation affecting the HERC2 gene is responsible for blue eyes. Professor Hans Eiberg and his team discovered that this mutation occurred between about 6,000 to 10,000 years ago.[24] These scientists argue that prior to this mutation, everyone had brown eyes. We know that some mutations produce diseases, and this is a very active field of research.

Many kinds of life have flourished on planet Earth during the past 3.5 billion years, and their particular molecule of DNA determined what kind of life they were. Many, including dinosaurs and early hominids, have gone extinct; they are no more. But we have fossil records to explain some of their history. Members of Homo sapiens sapiens are different from earlier kinds of life such as Homo erectus, Homo habilis, Denisovans, Neanderthals and others who inhabited the planet over the last few million years.

We may not yet have a formal definition of life that everyone can agree on, but we can say: If life is a symphony, then DNA is the conductor of the orchestra, and the music being played is just astonishing. People today dance across the same dusty old floors scuffed by previous generations, and many of the dance steps we take today were choreographed by the actions of those who lived before us. Their DNA was shaped by their actions, and our DNA has roots in theirs. In large measure, we are whatever we are because of who they were. The music plays on.

Out Relatives

The Crude and The Cultured

As mentioned above, two kinds of Homo sapiens have rambled about on the planet during the last 100,00 years or so. The first kind are called *anatomically modern* (the crude relatives) and the second kind are called *fully modern* (the cultured relatives). Our crude relatives were hunter-gatherers who roamed the African continent in small groups sometimes called tribes. Even though the descendants of these hunter-gatherers would eventually populate the Middle East, Asia, Europe and the rest of the habitable planet, their beginnings were meager. These folks were not farmers. In their search for calories, the tribe hunted animals and gathered whatever edible plants nature provided. They ate wild grains, tubers, nuts and fruits, and as they gathered calories for themselves, they had to avoid becoming calories for others.

They preyed on animals they killed themselves, and they may have also scavenged the remains of animals killed by other predators. Theirs was not a sedentary lifestyle. Hunting wild animals was physically demanding and very dangerous.

Large herds of grazing animals can exhaust the available forage in an area rather quickly, and when they do, the animals must migrate in order to find new sources of food. Not only so, but the early hominids could also exhaust the seasonal fruits, nuts, tubers and grains, and whatever else they gathered for food in their local area. Supplies in any locale were limited, and the larger the tribe, the quicker the food supply would have been exhausted. The anatomically modern folks lived in small tribes out of necessity. The San people and the Hadza people of Africa still live this way. The economics of this lifestyle prohibited tribes numbering in the hundreds, let alone in the thousands, even if they had the inclination to do so. We cannot be sure, but these early tribes of hunter-gatherers probably numbered less than a few dozen.

Our ancestors, the crude and the cultured, were not the only animals dependent on the migratory herds for calories, and this common need for food put our relatives in competition with some serious predators. Instead of always being the predator, members of our species also became the prey. It is difficult to imagine a lifestyle more laden with traumatic experiences — organized war, perhaps, but wars do not last thousands of years. Our ancestors lived with this extreme trauma for some hundreds of thousands of years. Still, with their limited cognitive abilities, the early relatives had no choice. Hunger is a powerful motivator, and if folks wanted to eat, they had to follow the herds and deal with the other predators as best they could. As is always the case, some of our relatives were more capable than others. In the competition for food, water, shelter and mating partners, physical prowess looms large. Not only so, but natural weapons such as sharp claws and strong teeth are important assets. Sharp eyesight, a keen sense of smell and good hearing are important, and it doesn't hurt to be able to run fast and climb trees to escape. So how did our early relatives rank in these categories?

Well, no human is as strong as an adult chimpanzee, let alone a lion or a leopard. Homo sapiens do not have the sharp claws or strong teeth as do the other predators, and compared to the eyesight of eagles and other birds of prey, we are virtually blind. We cannot escape danger by taking to the air as birds do, and humans are rather slow. The big cats and the other predators can easily catch even our fastest runners. Climbing a nearby tree to escape hyenas or wild dogs would have been helpful, but climbing a tree is pretty much useless when leopards or lions attack. Considering the huge disparities in strength, speed, weaponry, sense of smell and eyesight between human beings and the predators, it seems obvious that the only way we could have survived was by our wits.

What did our relatives know that the predators did not? In a word, fire. One of our relatives learned to control fire. No other animal on the planet can start a fire, and fortunately, the other animals are afraid of fire. This was true back then, and it is true today; fire is a powerful deterrent against wild animals. There were other reasons for our relatives to band together in tribes. Awful as it might be to point out, even if a local tribe was caught without a fire, a pack of predators can only eat a limited number of individuals at one meal. Those fleeing such an attack heard the screams of those unfortunates who were caught by the wild beasts. We imagine the escapees probably got busy making a fire right away.

The cognitive condition of our relatives was headed for a dramatic change, one of such magnitude that it hurled our crude relatives into a new era. Our crude relatives transitioned to our cultured relatives. This change took place sometime during the last hundred thousand years, and the change was not due to long term evolutionary refinements such as increasing brain size. In fact, on average, the Neanderthals had larger brains than Homo sapiens. Such a dramatic change is called *emergent*. As mentioned in the Introduction, an emergent event is one that is novel and surprising. It is a change that could not have been predicted by what came before, and it cannot be explained by some previous fundamental theory. For example, the birth of the cosmos was an emergent event. The appearance of life was emergent. So, what happened to change our crude relatives into cultured relatives?

Masters of the Planet is one of the most important books I have ever read. In it, Professor Ian Tattersall argues that some extraordinary, yet to be identified, cultural stimulus catapulted our ancestors from primitive to modern behavior.[25] This changeover was so dramatic that the only reason we have to believe something like this *could* happen is that it *did* happen. Professor Tattersall doesn't tell his reader what the event was, but he is right on the mark. Something extraordinary did happen, and the evidence suggests, at least to me, that this emergent event took place in the part of the world known as the Fertile Crescent.

My conjecture is the cultural stimulus Professor Tattersall argues must have taken place involved a special man named Adam. Whatever happened, it propelled our ancestors out of a lifestyle driven mostly by instinct, into one driven by symbolic thinking and complex language. Somehow our crude relatives became our cultured relatives.

Creatures of instinct are not moral agents; they just do whatever is necessary (and possible) in order to survive and to get what they need and want. Even though the emergent event changed the anatomically modern Homo sapiens into fully modern Homo sapiens, it was not a universal event that changed all kinds of life. It did not change the powerful predators into peaceful pets. The fully modern folks still had to deal with the predators, but following the changeover, they were much better equipped to do so. For the present, we continue by describing who our relatives were and how they lived.

When our relatives, those who already had the necessary body parts for modern behavior, realized that objects can have names, enormous new possibilities emerged. They no longer had to determine through personal experience what was edible and what was poisonous. They (we) could tell others where water and food could be found. We could plan hunting and gathering trips and warn others of danger. Any ability we may have had for tool-making could be explained as well as demonstrated. In short, we could tell our comrades where and how to get what they needed to survive, including how best to deal with predators. These new cognitive abilities also came with new liabilities. Fully modern Homo sapiens is the first species with the power to destroy all life on Earth.

If we don't wipe ourselves out with biological or nuclear weapons, or some other as yet unknown deadly technology; if we don't make the planet uninhabitable with noxious gases, discarded plastics or radioactive refuse; if we don't unleash artificially intelligent robots that turn on us; then maybe, just maybe, we'll surpass the 400,000 year reign of the Neanderthals. There are other contingencies we do not control that might wipe us out: A big solar flare, a nearby supernova, a large meteor or comet strike, extreme volcanic activity, or a wandering black hole that gets too close, for examples. Still, we remain optimistic that we have a few more years to explore our possibilities as fully modern Homo sapiens.

The cultural stimulus resulted in a huge leap forward for our species, and in the following, we will offer an explanation for how it came about. For the present, it is enough to note that our unique cognitive abilities are found in no other species. No other animal can start a fire or craft a bow and arrow or name objects, let alone build a rocket, fly to and land on the moon and return safely to Earth. The leap forward opened the door to astonishing new possibilities.

We know DNA determines physiology. It is important to understand that even though the potential (the necessary biological machinery) for symbolic thinking and complex language was put in place through changes in DNA, this potential for fully modern behavior lay fallow until some cultural stimulus sparked the changeover from non-symbolic behavior to the symbolic behavior we employ today.

Cultured Relatives

So, what was this culture stimulus? My conjecture is the stimulus that brought about the cognitive change in anatomically modern Homo sapiens was their interactions with Adam. Neanderthals and the anatomically modern Homo sapiens were hunters and gatherers, and as they roamed about searching for food, some of them almost certainly took notice of Adam. In their struggle for survival, the more primitive hominids had learned to be stealthy when the need arose. They may have watched Adam from concealed positions until they were convinced he was not a threat. Regardless of exactly how, or precisely when, the changeover from non-symbolic (anatomically modern) to symbolic (fully modern) behavior began, it was codified soon after Cain took a wife from among these other folks and taught them how to farm. Farming is prerequisite to building cities. Those who hunt and gather for their calories don't build permanent settlements. To be clear, my conjecture is that anatomically modern Homo sapiens came to understand that objects can have names through their interactions with Adam who was created with these abilities.

For reasons discussed further down, my best guess is the cultural stimulus took place between 100,000 to 23,000 years ago in the part of the world we call the Fertile Crescent.

PART III

Where Are We Going?

Preamble: Part III

In Part I we answered Gauguin's first question by noting that each person is made from trillions of atoms produced long ago inside stars. We really are star dust, as the late Carl Sagan was fond of saying.

In Part II we partially answered Gauguin's second question by noting that we are hybrid descendants of anatomically modern Homo sapiens who emigrated from Africa into the Fertile Crescent where some interbred with Neanderthals. We really are, at least in part, members of the long evolutionary line of hominids. This was a partial answer because it ignored the contribution of Adam and Eve, a deficiency we remedy in the subchapter "Diversity In The Garden." Whether it suits our Sitz im Leben or not, we came from the stars, and we are hybrids.

In Part III we address Gauguin's third question. When Gauguin asked, "Where are we going?" he was not asking where our bodies are going. There is no doubt on this point. The atoms that form each human being, from princes to paupers, are going to the same place. As it is written, *For dust you are, And to dust you shall return.*[26]

Gauguin asked where our soul goes when we die. According to scripture, at the moment of death, some souls will go to heaven, but others will go to hell. One or the other of these two things is going to happen to each of us in the relatively near future. Makes no difference if the person is a hero or a coward, a Nobel laureate or an idiot, when he or she dies, the atoms of their bodies will go off to do other things, and their soul will journey to heaven or to hell. The final destination of each soul will be determined by a decision each person will inevitably make, whether they purposely choose to do so or not.

Dogma

Dogma is understood as any scriptural truth claim considered by church authorities to be beyond question. The problem with this view is some of these dogmatic truth claims are beginning to buckle under the weight of modern science. The scriptures have remained fixed for centuries, but science has evolved and increasingly so during the last hundred years. Science is a two-edged sword. Swung one direction, it cuts away dogma that scientific scrutiny shows to be false. Swung the other direction, it cuts away unfounded criticisms of what the scripture claims to be true. In the following, we swing it both directions. As we compare and contrast the truth claims of science with the corresponding truth claims of scripture, it becomes clear that there are, in fact, certain dogma that are confirmed by science and other dogma that are refuted.

When science and scripture seem to clash, we must do two things: First, we must make sure we have properly understood what the truth claim of scripture actually is, for it is possible to misunderstand what some passages of scripture actually says. If we remain convinced that our understanding of some passage is correct, even though it conflicts with some well established tenet of science, then we must do the second thing. We turn to the evidence to determine whether scripture or science is correct. If the evidence requires it, we must either reformulate, or totally abandon the dogma that science clearly shows to be false. These are not trivial exercises, and they will certainly arouse the ire of some, for new ways of thinking about biblical dogma challenge our Sitz im Leben.

We discussed the Sitz im Leben in Part I, and how it adversely influenced Albert Einstein when it came to the proper interpretation of the equations of General Relativity which he authored. The Sitz im Leben is operative in all areas of human investigation, but its influence on scientific and scriptural matters is somewhat different. Its negative influence on scientific understanding, such as that seen in the Einstein episode, is often temporary because scientific evidence usually follows relatively quickly on the heels of new scientific ideas, and the evidence determines whether or not some new idea has merit.

Things are not quite as simple when it comes to scripture. Evidence that bears on the truth claims of scripture may be slow in coming and, in some cases, may never come forth. Faith, as a central tenet of the Christian religion, is like that. Even though those who trust the biblical truth claims do not require scientific evidence to sustain their faith, neither do they ignore evidence when it becomes available. God gave us our cognitive abilities, and God is not offended by honest questions. To the contrary, the Apostle James clearly states we are to take our questions to God.[27]

Long periods of time can pass before scientific evidence surfaces that allows a scriptural truth claim to be tested, and this has been our predicament for centuries, but no longer. We have entered a new and exciting age. Finally, we have scientific evidence whereby we can test a variety of scriptural truth claims. Notwithstanding, the Sitz im Leben is as persistent as one's shadow on a sunny day, and prudent investigators will be aware of their own.

For a simple example of the influence of the Sitz im Leben, consider the story of the three wise men who came to Bethlehem to worship baby Jesus. This story is widely known among Christians, and it is rarely scrutinized for accuracy. The second chapter of the Book of Matthew reports that men called Magi traveled from the East and brought gifts of gold, frankincense and myrrh to Joseph and Mary. As the story is told, since there were three gifts, there were three wise men. So, is this scenario accurate? Well, maybe, but maybe not. The Bible explicitly states that there were three gifts, but the Bible does not mention the number of Magi who made the trip. It is certainly possible there were only three, but given the hazards of travel in those days and the precious nature of their cargo, is it reasonable to believe men who actually were wise, and obviously wealthy, would make such a dangerous journey without an entourage that included what we would call, security guards?

Furthermore, even though the church gives short shrift to the value of the gifts, what kind of wise men would make such a journey in order to pay homage and then offer gifts consisting of only a smidgen of myrrh, a single gold coin or just a pinch of frankincense? Rather, and in contrast to the frequently emphasized poverty of Joseph and Mary, it seems more likely, at least to me, that the gifts of the Magi were substantial, if not lavish. In this view, the Holy Family was well provisioned for their trip to Egypt, a journey made necessary in order to prevent King Herod from killing their baby. However, whether or not Joseph and Mary had traveling money is not the point. This example is meant to illustrate how the Sitz im Leben can lead to tenuous interpretations that are not supported by the biblical texts. We have been conditioned to believe the Bible says there were three wise men, and these ruts are deep, but the number of Magi cannot be determined from the text. Maybe there were only three; maybe there were three dozen; maybe there were three hundred. Whether there were three or three hundred is not a matter of church dogma. We cite this simple example in order to demonstrate the power of the Sitz im Leben as it pertains to the interpretation of scripture.

Another, and perhaps more important example of the Sitz im Leben at work, concerns the timeline of creation. Those who claim the cosmos was created in six, 24-hour days, really do invite ridicule from those who understand the science discussed in Part I. We have solid theory and hard evidence which confirms the cosmos is about 13.8 billion years old, the Earth is about 4.5 billion years old, and life on Earth began about 3.5 billion years ago. Rome wasn't built in a day, and whether it suits one's Sitz im Leben or not, the cosmos was not created in six. If a Christian does cling to some dogma that scientific scrutiny shows to be false, this does not mean the person is narrow-minded, naive or stupid. It only means they are either unaware of the influence of their Sitz im Leben, or they are not gifted with the necessary scientific learning, as St. Thomas Aquinas so graciously put it. Despite what we have learned from science, some Christians still cling to the dogma that the cosmos was created in six, 24-hour days.

Old dogmas do not go quietly into the night, and quite frankly, this is a good thing, for in our enthusiasm to separate fact from fiction, we must be careful, lest we throw the baby out with the bathwater. The difficulties we encounter along the way are worth the effort, for if we cling to that which science clearly shows to be false, then we deserve the ridicule from the infidels. Not only so, but we risk making the worst of all possible decisions, the one that would prevent our soul from going to heaven when we die. Church dogma is not, nor should it be, immune from scientific scrutiny.

In order to objectively compare and contrast scripture with science, readers should begin by taking the Bible literally and seriously. Failure to do so is a biased approach that can foster fanciful interpretations which ignore what the texts actually say. If the Bible does not mean what it says, then the flood gate is flung open to even the silliest interpretations. However, if science shows that a literal reading of the Bible does not support our understanding of a particular passage, then we should honestly ask ourselves if we might be one of those St. Thomas Aquinas referred to as "not gifted" with the necessary scientific learning. On the other hand, if we do consider ourselves to be among the scientifically gifted, then perhaps our misunderstanding of scripture is rooted in some other area.

The Bible is primarily a book of history, but it also contains other genres such as poetry, metaphor, allegory, euphemism and so on. It could be that we have misunderstood a particular truth claim from scripture because we have failed to recognize the genre of the passage under consideration. On the other hand, we might have gotten the genre correct but then unwittingly allowed our Sitz im Leben to foreclose consideration of alternative interpretations. In this case we could say that our Sitz im Leben is shoving us into the shadows, rather than lifting us into the light.

Whatever the reason, if science shows that Christians have hung onto a mistaken idea about a biblical truth claim, then we are indebted to science for forcing us to re-examine some dogma which actually is false. If, on the other hand, atheists who have been rejecting some scriptural truth claim discover they were mistaken, then they owe scripture a debt. Whether either one acknowledges their debt to the other doesn't really matter. It's truth that matters. So, mindful of the different genres, we begin with a literal and serious reading of scripture, which we then compare and contrast with the applicable truth claims from science.

St. Thomas Aquinas was not the only person to recognize that religion and science are both important in our search for truth. For instance, some think Albert Einstein did not believe in God. Others think he did. For our purposes, it doesn't really matter one way or the other. Now that he is dead, he knows for certain that God is real, or, if God is not real, Einstein is beyond knowing anything at all. Whether Einstein believed in God or not, he did say, "Science without religion is lame; religion without science is blind."[28] So, in different ways, both Einstein, and Aquinas pointed out what some refuse to recognize today. Namely, we need both science *and* religion in our pursuit of truth.

As discussed in Part II, we have unequivocal evidence that several species of hominids lived on this planet for a few million years before fully modern Homo sapiens came along. All the others, including Neanderthals, are extinct. But Neanderthals left behind more than a few skeletons in caves. The evidence shows that Neanderthals and anatomically modern Homo sapiens exchanged genetic material in the relatively recent past, and due to these sexual encounters, modern people (outside Africa) carry some Neanderthal DNA in their genomes. These snippets of DNA from Neanderthals are rather like tiny skeletons in our genetic closets.

These scientific discoveries describe a history very different from the one imagined by the early commentators on the Bible. Unaware of the other hominids, no commentator could have made sense of certain parts of the Genesis narratives, which, by the way, editors divided at the wrong place when they inserted the Chapter Two heading. (Our earliest manuscripts of Genesis did not have chapter or verse designations.) So, through no fault of their own, the editors were driven to believe each of the first two chapters of Genesis tell a slightly different version of the story of Adam and Eve. After all, lacking any knowledge of Neanderthals or other primitive hominids, who else, other than Adam and Eve, could have been the males and females mentioned in the first chapter of Genesis? In all likelihood, this question never occurred to them. This single faulty assumption has done more to tie the first two chapters of Genesis into a Gordian Knot than any other issue we can name. As discussed in Part II, we now have a fairly good understanding (at least in outline) of early hominid history, and because we do, we are finally in position to explain a good deal of what has been so confusing in the past.

When first presented to the public, some the truths discovered through science challenged the Sitz im Leben of many. Even today, there are a few who are skeptical of the Big Bang and of evolution, but for the most part, even the strongest objectors to these scientific truth claims have pretty much settled down to reality. Their problem is how to reconcile what they have learned from science with what they read in the Bible. So, what do we know about this ancient religious document?

The Holy Bible

Although Bibles are bound into a single volume, each Bible is really a book of books, a total of 66 in the King James Version. The first 39 books are known as the Old Testament, and the remaining 27 are called the New Testament. Our oldest Biblical manuscripts were written in three languages, none of which were English. The Old Testament was written mostly in Hebrew along with a smidgen of Aramaic, and the New Testament comes down to us in Greek.

Some translations include additional books known as the Apocrypha. These additional books are understood to be important by those who acknowledge them, but they are generally considered (by Protestants) to not carry the same authority as the first 66. In addition, there is a significant body of other religious writings called pseudepigrapha. Many of these spurious documents claim to be written by mostly Jewish prophets. They may hold some historic interest, but they are roundly rejected as Holy Scripture. We are concerned primarily with the Book of Genesis — not with who wrote it, but with the truth claims it contains.

We cannot be certain about the authorship of Genesis because we do not have the original which, if Moses did write it, would be called the Mosaic autograph. Obviously, we cannot establish the date of what we do not have, and we can only work with what we do have. Consequently, we work with copies, and copies of copies, that were produced hundreds of years ago. Also obvious is the fact that those who cannot read Hebrew or Greek for themselves must rely on translations produced by those who can. Even though dating the Book of Genesis is beyond us, we have confidence that the New Testament books were written within a few decades after the death of Jesus of Nazareth. Although our primary focus will be the Book of Genesis, we will make occasional reference to other parts of the Old Testament and to the New Testament when appropriate.

The task before us is to compare and contrast the truth claims in the first six chapters of Genesis with the corresponding truth claims rooted in various areas of science such as the Big Bang and evolution. The evidence shows that Genesis includes key tenets of both these scientific theories. However, Genesis does not include the finer details of either.

If the evidence shows that some well-established truth claim of science conflicts with a truth claim of scripture, we will not sweep the matter under the rug. We are seeking truth, and we are committed to following the evidence, even if the evidence requires us to revise or totally abandon some dogma that has been part of our Sitz im Leben for many years. The Book of Genesis is our primary source, and it was not written in English.

Countless generations, regardless of their level of education, have found the Bible to be relevant and valuable. This is somewhat amazing in light of humanity's changing understanding of ourselves and of the world around us. How could any document written so long ago, in a time when people were scientifically ignorant, still be relevant and valuable? An atheist might say the Bible is not now, and never has been, relevant or valuable, but book sales tell a different story. Sales are the clearest indicator of which books readers find valuable, and for as long as annual book sales have been tabulated, the Holy Bible has outsold any other book. Whether one is delighted, depressed or indifferent, the facts are clear: Readers have voted with their dollars, and the results show beyond doubt that the Bible is still a relevant and valuable resource. So how can this old document still be relevant in the age of science?

Well, the inherent value of any document is determined by its content, and modern people have become accustomed to intelligent women and men figuring out how things work and sharing their insights with the rest of us. The style of writing is important, but content is the central issue. These leaps in scientific understanding are the products of the human mind, one most often prepared by a proper education and exercised through diligent work. Those who make these leaps in understanding deserve credit for their work, and we reward these luminaries with accolades commensurate with their achievements.

Rev. John Wesley, founder of the Methodist Church, believed the Bible is the most important book ever written, for in it, God teaches the way to heaven. As Wesley said, "God himself has condescended to teach the way. He hath written it down in a book." The Bible is precious because the content is from God. Even so, those who seek Biblical truth have some challenges to overcome.

Conflicting Truth Claims

Even though God provided the content, debate continues on certain issues because careful reading of any English translation one might choose reveals a number of inconsistencies between the first two chapters of Genesis. These inconsistencies cannot be resolved by wishful thinking, special pleading or theological gymnastics, as is evident in numerous failed attempts to do so.

Inconsistencies foster doubt, not only about the trustworthiness of the passage under consideration, but also about the remainder of the Bible. If Genesis cannot be trusted to be accurate, why should the reader believe the rest of the Bible? Skeptics of the Bible have a legitimate question here, and emotional answers that ignore scientific evidence do not help; we need more light, not more heat. Defenders of the Bible should offer gentle answers to legitimate questions and certainly should abstain from harsh answers meant to castigate skeptics.[29] After all, the skeptics didn't write Genesis; in the story I'm telling, Moses did. And we, along with New Testament authors, are claiming that Moses wrote under the inspiration of the Holy Spirit. And, what's more, *God is not the author of confusion.*[30]

These are not new issues as can be seen in the New Testament guidance on how to respond to legitimate questions that arose in those times. In the following, we parse these inconsistencies according to the relevant evidence we have gleaned from science.

Since God is not the author of confusion, the inconsistencies in the early parts of Genesis came about through the work of editors. They introduced the inconsistencies in their efforts to harmonize the conflicting truth claims they saw, primarily between the story of the males and females in the first chapter of Genesis with the story of Adam which begins in Chapter Two. The inconsistencies they created went pretty much unnoticed and unchallenged until the growing body of science began to bring them into the light. In this view, the editors took on an impossible task because the first two chapters of Genesis recount two very different sagas of history.

Chapter One of Genesis tells the story of God initiating processes that brought forth three kinds of primitive life which evolved over the past 3.5 billion years into the diversity of living creatures we see around us today. The evolutionary stream of life reached its apex in the form of anatomically modern Homo sapiens who show up in the fossil record around 300,000 years ago. As noted in Part II, the evidence shows these hominids evolved in Africa, and some of them immigrated into the Middle East where they produced hybrid children with the Neanderthals.

In sharp contrast, Chapter Two of Genesis recounts the story of God's one-off creation of Adam and the subsequent cloning of Eve. Adam and Eve did not arrive on the stage of history through evolution. To be clear, my conjecture is that Chapter One of Genesis makes no mention of Adam or Eve. Well intentioned, but poorly informed, editors are the authors of these confusions.

The early editors of Genesis were academics of their day, scholarly men who brought to their desks not only special expertise in the Hebrew language and theology, but also a profound reverence for the manuscripts. They would have been very reluctant to substantially change any of the content of the manuscripts and certainly would not have changed the quotes attributed to God. After all, who would have the audacity to put words in God's mouth? So we proceed on the assumption that the quotes attributed to God were not changed in the slightest, but the commentary of Moses might have been jiggered about as the editors sought to "fix" the inconsistencies they believed they saw in the texts. Even so, the Mosaic commentary was, at least for the most part, unchanged — an assertion we can substantiate by comparing the truth claims found in Genesis with the truth claims of modern science. If the ancient truth claims in Genesis correspond with the truth claims of modern science, then the early editors could not have significantly changed the content of the originals. If they had, then the claims of scripture would *not* agree with those of science. Ironically, as mentioned earlier, the early editors deemed their work necessary because they did not have the scientific information required to understand the texts in the form they had before them, and because we now do have the science, it should be possible to identify their edits and work out how the earliest texts most probably read.

The discovery that the universe had a beginning is arguably the most important scientific discovery of all time. And we can say that the second most important discovery is what we call evolution. The nomadic Hebrews would not have understood the details of either of these discoveries, and literature that cannot be understood has little value to its reader. If Genesis had been written in the scientific jargon used by professionals today, then instead of being highly valued and carefully copied over many generations, it would have rather quickly ended up in the dust bin of history. For the Bible to have made sense to scientifically ignorant people, which it clearly did, and to millions of modern people, which it clearly does, is remarkable. And we have the sciences of the Big Bang and of evolution to thank for new insights into the history recorded in the Book of Genesis.

Genesis

Some of the Challenges

All truth passes through three stages. First, it is ridiculed. Second, it is violently opposed. Third, it is accepted as being self-evident.

— Arthur Schopenhauer (1788 - 1860)

Schopenhauer was awarded a doctorate from the University of Jena. His dissertation, *The Fourfold Root of the Principle of Sufficient Reason,* formed the centerpiece of his philosophy. Some of his new ideas faced challenging hurdles which may have prompted the above aphorism.

A commentary on Genesis is a collection of opinions which the author believes best explains the meaning of the text. Commentaries can be the work of an individual or of a group, and scholarly commentaries often include explanations of the analysis that underpin the opinions expressed. They also evince consideration of problematic words and grammar, along with alternative interpretations. Authors of commentaries are expected to have special expertise in the subject matter and to write from an unbiased perspective. Attaining the special expertise is more easily accomplished than is writing from an unbiased perspective. Like a commentary on any other text, the opinions expressed, even by experts, will be influenced by the Sitz im Leben of the author(s) of any commentary on Genesis. This is not to say all opinions are incorrect; certainly, they are not. But neither does it mean all opinions are correct. When opinions differ, we turn to evidence to decide the matter. Surely, each author expresses what they believe to be correct.

Here's our primary challenge: As traditionally interpreted, the first two chapters of the Book of Genesis make conflicting truth claims about the origin and history of our species. They just do. And pretending these issues do not exist, not only provides fodder for the critics, but also makes it difficult for the faithful to know what to believe. For example, in Chapter One we read that God created man in His image, created them as male and female, and gave them dominion over the other animals. God instructed these males and females to be fruitful and multiply and to fill the earth with their children. There is *nothing* in this passage to suggest these early hominids had the ability to speak, *nothing* to suggest they engaged in agriculture, and *nothing* to suggest they had souls. Therefore, anyone who assigns these three attributes to the males and females of the first chapter of Genesis does so because of their Sitz im Leben, not because of what the text actually says.

In stark contrast, we read in Chapter Two of Genesis that God formed Adam from the dust of the earth, and breathed the breath of life into Adam's nostrils, whereupon Adam became a living soul.[31] Furthermore, God put Adam in the Garden of Eden to till it and keep it. Obviously, Adam had the mental wherewithal to engage in agriculture. Subsequently, God brought the animals to the garden for Adam to name them, and whatever Adam named them, that was their name. This naming episode unequivocally shows that Adam had the cognitive ability for symbolic thinking and language. Furthermore, Adam was a bachelor until God cloned Eve from one of Adam's ribs, and bachelors, at least those without sexual partners, don't produce enough children to populate a small garden, let alone the entire planet.

Traditionally, these two chapters have been understood to recount different versions of one story whose main character was Adam. These are deep roots, and any alternative explanation will certainly challenge the Sitz im Leben of those who believe the first two chapters of Genesis really are two versions of one story. Sitz im Leben aside, they are not two versions of one story. Instead, they each tell a unique story. What appears to be conflicting truth claims in Genesis are rooted in the failure to appreciate the different sagas the first two chapters were written to address.

Fortunately, this problem of these conflicting truth claims can be solved by directly comparing various scientific truth claims with the corresponding biblical truth claims. We live in the age of science, and failure to show the correspondence between science and scripture leaves those interested in Genesis in a quandary. There are eloquent ways to frame the question at the root of all this, but we can reduce the field to one simple question: Is Genesis fact or fiction? This problem is particularly pervasive among younger readers whose public educational experiences steeped them in science but starved them of scripture. Thank God for all the attention to science provided through public education, and God forgive us for the lack of attention to scripture. Failure to demonstrate the connections between science and scripture inevitably results in an unwarranted mixing of mythology with biblical history, and this has had a drastic and debilitating effect on modern society.

The late Reverend Dr. Martin Luther King, Jr., clearly stated our predicament when he said, "We have guided missiles and misguided men." If the problems discussed just above could be solved by demonstrating the correspondence of science with scripture, then why do we not already have such a treatise?

Well, the main reason for this poverty is the required science has only recently become available, and obviously, two things cannot be compared when one of them is missing. Without the science, it was impossible for early commentators to make the connections. For another thing, even today, the breadth of scientific knowledge required to demonstrate the connections is extensive, and this raises other challenging issues.

Nobel laureate Richard P. Feynman (1918-1988) acknowledged these difficulties during the first of three lectures he gave in 1963. He said, "In talking about the impact of ideas in one field on ideas on another field, one is always apt to make a fool of oneself. In these days of specialization there are too few people who have such a deep understanding of two departments of our knowledge that they do not make fools of themselves in one or the other."[32]

His observation is as applicable today as it was back then, and this author is well aware of this pitfall. Along the same line, the late Stephen Hawking pointed out that the rapid advances in all fields of science in these times make it impossible for any one person to be on the cutting edge of all scientific disciplines. Even so, our situation is far from hopeless. One need not be on the cutting edge of all scientific disciplines in order to explain many of the connections between science and scripture. Still, one must have a fairly good understanding of the scientific disciplines of interest. As Einstein said, "If you can't explain it simply, you don't understand it well enough." If one does not understand the science well enough to explain it simply, the task of comparing and contrasting the science with scripture is daunting, if not impossible. Even though we've had the scripture for millennia, most of the science required to objectively compare and contrast the two has only become available during the last few decades.

In light of Albert Einstein's, Stephen Hawking's and Richard Feynman's observations, we can say, not only must those who write sensibly about these topics have some knowledge of the Hebrew language, of various ambiguous words and phrases that might be misunderstood, but in addition, they must be familiar with the major tenets of the Big Bang, of evolution, of genetics and archaeolog. Failure to understand the science, or simply ignoring science will lead to translations of scripture which are clearly refuted by the very science ignored. As mentioned above, knowledge from these scientific disciplines was not available to early editors, so they had no chance of connecting the relevant science with the truth claims of scripture. Thankfully, we now do have the science required to cut through the Gordian Knot that has flummoxed the faithful in the past.

A word of caution is in order. There are some scriptural truth claims that have no scientific corollaries. In the following, we compare and contrast those scriptural claims that do have scientific corollaries. The other claims will have to wait.

The Challenges of Biblical Criticism

Over the centuries, scholars made attempts to explain inconsistencies they observed in the Holy Bible, particularly in the first five books of the Old Testament, which are called the Pentateuch. Along the way they developed a line of investigation known as biblical criticism. This term is offensive to some because the word "criticism" is often taken as only negative, as something intended to demean and devalue whatever is criticized. But this is not necessarily the case with biblical criticism. For our purposes, biblical criticism is the analysis of the Bible from a historical, textual and philological perspective, and biblical criticism can be positive, as well as negative.

These inquiries are important, but not as important as is bringing the bright light of science to bear on the scriptures, for if science concurs with what the Bible actually says, then whatever biblical criticism brings to the table doesn't really matter very much. The following is a very brief outline of the development of biblical criticism.

J. G. Eichhorn (1752-1827) is considered to be the father of what is known as higher criticism of the Bible. In his 1779 book, *Die Urgeschichte*, he argued that Moses is not the author of Genesis, but instead, editors combined multiple sources to produce what we have today.

Julius Wellhausen (1844-1918) was another famous critic. He is well known in scholarly circles for his *Prolegomena zur Geschichte Israels*. He also formalized what is known as the *documentary hypothesis* which asserts that Genesis is a compilation of materials from four different sources. Wellhausen believed he could identify the imaginary sources, which he called J, E, P and D depending in large part on the name each of the sources used for God.

Herman Gunkel (1862-1932) was another proponent of biblical criticism. He also believed that the Genesis narratives are derived from multiple sources. He lamented, "...the last great genius who might have created out of the separate stories a great whole, a real 'Israelitic national epic' never came." Gunkel believed that modern investigators are fortunate that this final editor never came along since his absence left the earlier legends intact, and thus, it is possible to discern the history of the entire process. He cautioned that, "...theologians should learn that Genesis is not to be understood without the aid of the proper methods for the study of legends."

Whether one approves of the various conclusions reached by the biblical critics or not, we are indebted to these scholars for helping us identify, evaluate and address certain inconsistencies within the Genesis narrative. For the sake of argument, let's assume for a moment that these biblical critics were correct, and see where the evidence takes us.

If Genesis actually is a compilation of various sources, as the biblical critics contend, then, at the very least, the final edition of Genesis was influenced by the Sitz im Leben of those who compiled it. And, mark this well, the Sitz im Leben of each person depends not only on what the person knows, but also — and even more importantly — on what the person does *not* know. In addition, the fact is that one does not know the extent of what one does not know! So what might have been some of the factors in their Sitz im Leben that would have conditioned their editorial choices?

Well, to their credit, the early editors would have had a more robust knowledge of ancient languages than do some modern translators. Not only were the early editors more familiar with the meaning of some ancient Hebrew words than we are today, they may have had a higher degree of reverence for the biblical text than some modern commentators. They believed scripture to be a vessel of divine communication that not only told the history of the heavens and the earth, but also the history of their nation. Parts of this history were a source of pride due to the noble behavior of those mentioned, and some of it must have been a source of embarrassment due to the shameful behavior of others.

Furthermore, if the inconsistencies they thought they saw in the texts were embarrassing, then their Sitz im Leben would have tended toward emphasis of the positive aspects of their history and de-emphasis of the negative. They almost certainly wanted to harmonize the first two chapters of Genesis.

We may not care to admit it, but human beings are inherently prideful creatures. We usually try to hide this penchant for pride from our peers, but even in the best cases, it lurks just below the surface in all but the most holy and humble members of our species. These early editors labored under the same handicap, but their pride was no match for their reverence for scripture, a reverence that forbade them from completely excising the embarrassing episodes.

Even though pride very likely played some role in their choices, it was not the most damaging element to their work. We are assuming a good deal and imagining more, but if the biblical critics are correct, one thing is clear enough; the early editors did their work without the scientific information required to do a thorough job of it. In particular, the editors did not know that several different species of hominids lived on this planet over the last few million years, and without this crucial piece of information, their options on how to understand the first two chapters of Genesis were limited.

Unaware of the complexity of hominid history, they were drawn into the mistaken idea that Adam and Eve were the male and female mentioned in the first chapter of Genesis, but nothing of

Adam and Eve can be found in chapter one. Instead, chapter one describes the evolutionary line of hominids, most of whom lived long before Adam and Eve stepped onto the stage of history. This is why the story of the males and females in chapter one is so very different from the story of Adam and Eve in chapter two.

To tidy up just a bit, I am arguing that the first chapter of Genesis describes the creation of the cosmos, the formation of the Earth and stars, the evolution of plant and animal life, and the evolution of early hominids who lacked the cognitive abilities necessary for symbolic thinking or agriculture. The second chapter of Genesis sets forth the history of Adam and Eve, who had both. The complexities and contingencies of hominid history continue in the third and fourth chapters of Genesis, for it is here Adam and Eve begin to interact with the evolutionary hominids.

The Challenge of Authorship

Aware of the arguments from biblical critics, but lacking any unequivocal evidence to the contrary, we follow the long-standing tradition that Moses was the author of Genesis. Jesus repeatedly referred to Moses when he spoke to Old Testament concepts.[33] For the sake of convenience, we will do likewise. Obviously, those who ascribe to the documentary hypothesis disagree. Even so, the task before us is not to argue the authorship of Genesis.

A serious challenge to our work is not having the original document, if there ever was one. Considering the extreme reverence for scripture within the Hebrew community, it is puzzling how this foundational document could have disappeared. The fact that we do not have the autograph lends support to the view of the biblical critics who argue there never was such a document. However, our not having the autograph does not settle the matter. There are plausible reasons for this poverty.

We know that foreign armies have waged war against Israel on numerous occasions; Assyria, Babylon, Egypt and Rome come to mind. Muslim armies overran most of the Middle East, including Israel, and ended Christian rule in the city of Constantinople in the Middle Ages. A coalition of Muslim countries instigated war against Israel in 1967. The invading armies were defeated in six days. Even today, there are groups who deny Israel's right to exist as a nation. So, perhaps the autograph of Moses was intentionally destroyed by those who sought to invalidate the historical claims of the Jewish people to the Holy Land.

Perhaps the Mosaic autograph was accidentally destroyed in some catastrophic fire, flood or some other natural disaster. Maybe it resides on a shelf in some library of other nondescript manuscripts. Maybe it remains buried in an unexplored cave in the Dead Sea region of Israel. Perhaps the Book of Genesis really is no more than a compendium of sources, gathered, redacted and compiled into its present form as Wellhausen and other biblical critics proposed. Such reasons are speculation, but, in fact, we do not have the original. So, we work with copies and whether we like it or not, copyists and translators make mistakes from time to time.

Along the way, we will point out what we consider to be emendations that have materially altered the meaning of the text. Some believe that Moses wrote the Book of Genesis, and others believe Genesis is the work of editors who compiled different sources. My best guess is that Moses wrote the initial corpus based on the information God provided, and with the passage of time, copies were created and circulated in different communities. If scholars edited, emended and compiled different copies into what would eventually become the modern versions we have today, they labored without the information required to understand the evolutionary aspects of hominid history, and this poverty led them to make some changes in the text that cannot stand the light of scientific scrutiny. In this view, the question of who wrote Genesis cannot be answered in an either/or manner. Even so, as mentioned above, the truth claims are whatever they are, regardless of who penned them.

Even though the following may resemble a new commentary on the first six chapters of Genesis, this is not my intent. Instead, the following is written to compare and contrast the biblical truth claims set forth in the opening chapters of Genesis with the corresponding truth claims of science. Notwithstanding, Genesis makes a small, but very important, number of truth claims that are couched in metaphor and allegory. In these cases, I will offer my humble opinion on how these metaphors and allegories should be interpreted. Obviously, this does fall into the category of commentary. Still, this is not a commentary, per se.

The first chapter of Genesis describes historical events that took place billions of years before Moses was born. This is astonishing! Somehow, he accurately described what we now understand as cosmological history, and these historical facts were literally impossible for him to know, at least through ordinary means.

Translation Challenges

Obviously, we can only work from what we have, and, as discussed above, our first difficulty is that we do not have the original of the book of Genesis. Without the original, we cannot even say with absolute certainty that Genesis was first written in Hebrew. We assume it was, and we work from this perspective, but from his infancy, Moses was raised by the daughter of Pharaoh. He was surely fluent in the language of Egypt. However, he also very likely knew Hebrew as well.

One of the challenges faced by modern translators is the meanings of some ancient Hebrew words have become uncertain over the centuries. In such cases, translators can only offer the reader an educated guess at the meaning of some words — not an altogether satisfying happenstance, but an unavoidable one. Herein is yet another instance, and a very crucial one, in which the Sitz im Leben of the translator looms large. The influence of the Sitz im Leben on educated guesses is but one difficulty we face when the meaning of some old Hebrew word is uncertain.

There are fundamental differences between Hebrew and English. Hebrew is a consonantal language, which means the alphabet contains no vowels, and without the phonemes (sounds) associated with vowels, pronunciation of the words is problematic. A group of scribes known as the Masoretes addressed this problem by developing a system of marks called vowel points sometime after about 600 AD. The vowel points serve to inform the reader of the correct pronunciation of the words. If there ever was a Mosaic autograph of Genesis, it almost certainly did not include vowel points, which means the Sitz im Leben of the Masoretes had some influence on the choices they made. There are other differences between the English and Hebrew.

The letters of the English alphabet are abstractions, which means any resemblance of an English letter to any object in nature is purely coincidental. In contrast, the letters of the Hebrew alphabet are pictographs, which means each one is a pictorial representations of some physical object the Hebrews knew quite well. For example, the first letter of the Hebrew alphabet, א (pronounced aleph) is a pictograph of the head of an ox. Oxen were known for their strength. So, when the aleph was used in the construction of some Hebrew word, not only did it carry the A sound, it also carried the idea of strength. Likewise, the Hebrew letter, ל (pronounced lamed), is the pictograph of the shepherd's staff; it carries the L sound along with the notion of authority or leadership. So the Hebrew language is much richer than the English.

Here's a difficulty that is rarely, if ever, mentioned: Most of the Genesis narrative is already commentary. It's the commentary of Moses. Inspired commentary to be sure, but commentary nevertheless. What's more, Moses, like any other human being, was somewhat constrained in what he was able to write. Yes, he was inspired to write by the Holy Spirit, but he had to use words that were part of his vocabulary. What we mean is this: If the Spirit had given Moses a scientific account of the particulars of the Big Bang, had given him the equations of general relativity and quantum mechanics and of the genetic code that drove evolution, it would simply have been beyond him to write about it in a manner that he could have understood, let alone his readers. Remarkably, the commentary of Moses was understandable and relevant to readers back then, and it is still relevant to some hundred million readers today.

We are about to shine the bright light of science onto the "Sacred Pages of Scripture." As we do, the Sitz im Leben will undoubtedly spring into action, but sooner or later, the evidence will determine whether our efforts will be accepted or rejected.

Truth Claims In Genesis

Four stages of acceptance:

i) this is worthless nonsense

ii) this is an interesting, but perverse point of view

iii) this is true, but quite unimportant

iv) I always said so

— J. B. S. Haldane

English geneticist J. B. S. Haldane (1892-1964) described four stages new ideas go through before they are widely accepted as common knowledge. His view is a bit different from that of Arthur Schopenhauer mentioned earlier, but the two agree that new ideas, even those that turn out to be correct, are bound to face serious challenges before they are accepted as common knowledge.

In order for us to grow in our understanding, we very much need new ideas. However, not every new idea turns out to be correct. Some ideas that are correct can be so powerful they hurl us forward in our understanding, and this inevitably leaves some of our older ideas on the ash heap of history.

For examples: Carl Woese taught us there were three kinds of primordial life instead of two; Charles Darwin introduced us to evolution; Albert Einstein taught us about relativity, and Reverend George LeMaitre introduced us to the Big Bang. We need new ideas, but according to J. B. S. Haldane, all new ideas go through stages of acceptance. Furthermore, some new ideas really do turn out to be worthless nonsense.

Failure to connect the truth claims of scripture with those of science leaves thirsty souls in a spiritual desert. Wandering about in this arid landscape, some abandon the literal meaning of the scripture altogether. Others just sweep contradictory claims under the rug and stroll along hoping things will somehow turn out okay. Those unwilling to fully abandon scripture had little choice other than to retreat into the land of mythology, a land where the Sitz im Leben looms large, and the ordinary meaning of words are frequently ignored. In this scenario, the texts can be interpreted to mean almost anything. This approach may foster entertaining theological theater, but it inevitably strips the spiritual sustenance from the scripture. We can say thirsty souls in this spiritual desert wander among the serpents as they struggle with what is fact and what is fiction.

Scripture and science make truth claims about many things, and in those cases where they make truth claims on the same topic, the two claims are complementary, not contradictory. In order to test this thesis, we begin by comparing the first two verses of Genesis with what we have learned from science. If the evidence shows that the truth claims in the first verses of Genesis actually are synonymous with the truth claims of science, then we have good reason to continue this line of investigation.

Truth Claims In The Headline

Genesis 1:1-2 is a headline for the creation narratives that follow it, and fortunately for us, this headline describes a number of key events which we now know took place during roughly the first 380,000 years of the Big Bang. These events are confirmed by solid theory and good evidence from science.

Albert Einstein started the cosmological revolution with his equations of relativity. Rev. George LeMaitre solved the equations and gave us the theory of the Big Bang. Astonishingly, the work of these two brilliant scientists is circumscribed in the first word in the Bible. Here are two renditions of the Hebrew word.

ברסשית ב ראש ית

The larger marks (the letters) that form these two words are consonants The smaller marks that appear below, above and inside some of the consonants on the left are known as *diacritical marks* or *vowel points*. The vowel points were added by later editors to help the reader know how the words should be pronounced.

Hebrew is written from right to left, so the first letter of this word is, ב (pronounced bah). It is a preposition translated as 'In'. The remainder of this word, רסשית (pronounced ray-**sheeth**), is translated as 'beginning'. Together, these two become "*In the beginning.*" Genesis, as the title of this book, derives from this word.

Our earliest manuscripts do not have the vowel points or divisions of the text into numbered verses. We have come to expect modern translations of Genesis to include numbered verses, capitalization, paragraph indentions and punctuation marks, even though they were later additions. We very much need and appreciate these editorial additions, but they were *not* part of our earliest copies of Genesis.

Translations of The Bible

Unless noted otherwise, all citations of scripture in the following are from the New King James Version (NKJV). This means that when the reader encounters some passage of scripture, it will be from the NKJV — unless the passage is noted to come from a different version of the Bible.

For convenient comparison, five different English translations of Genesis 1:1-2 are reprinted just below. They are denoted by the following abbreviations: the King James Version (KJV); the New King James Version (NKJV); the Revised Standard Version (RSV); the New Revised Standard Version (NRSV) and the New International Version (NIV).

In the beginning God created the heaven and the earth. And the earth was without form, and void; and darkness was upon the face of the deep. And the Spirit of God moved upon the face of the waters. (KJV)

In the beginning God created the heavens and the earth. The earth was without form, and void; and darkness was on the face of the deep. And the Spirit of God was hovering over the face of the waters. (NKJV)

In the beginning God created the heavens and the earth. The earth was without form and void, and darkness was upon the face of the deep; and the Spirit of God was moving over the face of the waters. (RSV)

In the beginning when God created the heavens and the earth, the earth was a formless void and darkness covered the face of the deep, while a wind from God swept over the face of the waters. (NRSV)

In the beginning God created the heavens and the earth. Now the earth was formless and empty, darkness was over the surface of the deep, and the Spirit of God was hovering over the waters. (NIV)

These translations were produced by scholars separated by decades in most cases, and by centuries in the case of the King James Version. So we should not be surprised that they chose slightly different wording. In most, but not all cases, slight variations in the different translations will not drastically change the meaning of the text or hinder our work. However, there are a few translation choices that are refuted by science. In the following, we address these exceptions as they arise.

The wealth of information contained in this headline is just astonishing, for with less than 40 words, it describes events which modern science confirms took place during the first 380,000 years of cosmic history. Even though the translations cited above use slightly different wording, each one begins with the same truth claim: namely, the universe had a beginning. We'll start our comparisons with this claim.

There is no single Hebrew word for universe. In biblical parlance, *the heavens and the earth* is the equivalent, so the first truth claim in Genesis is that the universe had a beginning. As explained in Part I, the answer from science on the question of whether or not the universe had a beginning has, during the past century, changed from a firm "no" to an unequivocal "yes". The fact that the cosmos had a beginning is arguably the most important scientific discovery ever made. Contrary to what Albert Einstein first believed, and Fred Hoyle always believed, the universe is *not* eternal. This very important question is settled, once for all, and science and scripture agree — not surprising since one truth will not contradict another truth. This truth claim from scripture has been sailing along on the sea of knowledge for thousands of years, and the ship of science has recently come alongside. Whether written in Hebrew scripture or in English books of science, the truth claim is the same: The universe had a beginning.

The second truth claim in the above passage of scripture is God created the universe. Even though science is silent on this claim, a number of scientists are not. Some scientists who are atheists offer speculations to try and explain the sudden appearance of the universe, but no way to test their ideas and no evidence to support their rhetoric. Jews and Christians believe God created the heavens and the earth, just as this verse claims.

The third truth claim in the passage above concerns the planet we inhabit, and this claim comes in two parts. The first part of the claim is: *the earth was*, which means the earth existed. The second part of the claim is: although the earth existed, it was *without form and void*. As explained in Part I, everything that now exists was, in some form or another, contained within the young universe. So the earth, or at the very least, the materials that would eventually become the earth, was part of the young universe. Stated differently, the earth really *was,* as Genesis claims. And, what's more, because everything in the young universe was in the form of plasma, the earth really was *without form and void*. Science confirms both parts of the third truth claim described above.

The fourth truth claim in this passage is that *darkness was on the face of the deep*. This sentence contains three keywords: *darkness, face* and *deep*. Continuing the tradition that Moses was the author, we can say that each of these words tells us something about what he saw. Moses chose the word "deep" to describe this entity, and whatever it was, he wrote that it had a "face", which means it had an outer boundary which we would call a surface. In addition, this surface he saw was "dark." So what was this entity that Moses described with these words?

My conjecture is that Moses saw the young universe from God's perspective, which means from the outside. Yes, that's correct; as we saw in Part I, there was no outside, per se, but we can be sure that God did not create Himself when He created the universe. God surely had an outsider view, and the description Moses penned strongly suggests that he was allowed to see the young universe from God's perspective. For the sake of argument, let's suppose Moses had a view of the young universe from God's perspective. Can science provide any insight on what one might expect to see from the outside? Well, yes, it can, and the three keywords Moses chose (*deep*, *dark* and *face*) tell us a great deal.

The deepest thing we can name in the physical realm is the universe itself. Terms such as "deep space" have entered the vernacular, and we understand that nothing is deeper than space. Furthermore, this observation is true at any point in time one might choose. Just as space is the deepest thing we can name today, it was also the deepest when it was one billion years old, and it was the deepest when it was one minute old. Even so, the young universe was not infinite; it had an outer boundary, a surface. Moses called this outer boundary "the face".

As pointed out in Part I, we have learned through the science associated with the theory of the Big Bang that even though the cosmos was overwhelmingly made of light, for about the first 380,000 years the universe was dark because it was in the form of plasma, and the plasma was opaque to the astonishing light within. Solid theory and multiple streams of scientific evidence show this to have been the case. Without resorting to the jargon associated with the Big Bang, it would, even today, be difficult to write a better description of what the young universe looked like from an outsider's perspective. It was *deep*. It had a *face*, and that face was *dark*. Thus, with regard to the early universe (*the heavens and the earth*) the truth claims of scripture and those of science are synonymous, and because they are, the idea that scripture is nothing more than a collection of mythological tales must be rejected.

The complementary nature of science and scripture in the above examples shows that Moses described things that were utterly impossible for any human being to know at the time, at least through natural means. Obviously, the events he described took place billions of years before any human being was around, and yet, they correspond very closely with what we have learned from modern science. Thus far, our comparisons have been straightforward and easy to understand, but things get a bit more complicated in the last part of verse two.

In Genesis 1:2c, the KJV, NKJV, RSV and NIV each report "the Spirit of God" was doing something at this early stage of cosmic evolution. In contrast, the NRSV reports "a wind from God" was doing something. The NRSV translation of this verse is refuted by science; here's why. The Hebrew word, רוּחַ (pronounced **roo** - akh), can mean wind or spirit or both, and the context usually makes clear which meaning is appropriate. So which of these translation choices fits best with what we know from science? Well, unless we rob the word of its ordinary meaning, wind denotes movement of the atmosphere above the surface of the Earth, and as explained in Part I, there was no recognizable Earth at this early stage of cosmic expansion. Yes, the Earth *was*, but it was *without form and void*. The expanding cosmos had a *face*, but the Earth did not.

Furthermore, there was no atmosphere. The Earth's atmosphere today is about 80% nitrogen and about 19% oxygen, along with lesser amounts of other gases. As explained in Part I, nitrogen and oxygen were formed inside stars some hundreds of millions of years after the beginning. Since the Earth was *without form and void*, there was no solid surface for the wind to sweep over, and there was no atmosphere anyway. So on the basis of well-established science, as well as the availability of better translation choices found in most other versions of the Holy Bible, *wind*, as a translation choice, must be rejected. The proper translation of the word in question is, *Spirit*, not *wind*. According to Robert C. Dentan, the NRSV committee members chose, wind because they saw no creative function ascribed to the *ruach*.[34] (In fairness, he pointed out that not everyone on the committee agreed with this translation choice.) Genesis 1:2c speaks of the Spirit of God.

As noted previously, Albert Einstein said, "Science without religion is lame; religion without science is blind." If one is unaware of the science, or does not understand the science, or chooses to ignore the science, then certain translations will be refuted by the science. I did a cursory internet search to see if the NRSV committee members had scientific credentials. As expected, they were highly respected scholars in their fields, but I did not see any indication that physicists were on the committee. However, it was only a cursory search.

Ignoring the science fosters poor translation choices. Moreover, failure to recognize the importance of another Hebrew word adds to the problem. The word is רָחַף (pronounced raw-**khaf**). This is the Hebrew word one would use to describe a bird hovering over her young. Birds are indeed doing something creative when they hover over their young: preparing to mend the nest or protecting the chicks or preparing to feed them or demonstrating how to fly or some other thing only the birds would know. Something (actually, Someone) was hovering above the expanding hypersphere of plasma, and the NIV and the NKJV provide better descriptions of what was happening. They translate this verse as *the Spirit of God was hovering over the face of the waters,* and they have captured the moment. The Spirit of God was *hovering over the face,* but we have yet another problem in so much that *the Spirit* was not hovering over the face of *the waters.* This could not have been the case. Science shows there was no water at this early stage in the evolution of the universe.

All the versions listed above translate the Hebrew word, מַיִם (pronounced **mah**-yim) as water, and even though other translation choices are known, water is the consensus choice. The problem, based solely on well established science is this: just as there was no wind, neither was there any water at this early stage of cosmic evolution. As discussed in Part I, the early universe was a boiling caldron of plasma, and because it was, there was no water. As is well understood, water is formed of two atoms of hydrogen and one atom of oxygen (H_2O), neither of which was present during the epoch this passage of scripture addresses. The early universe contained lots of protons and electrons which later combined to form atoms of hydrogen, but until atoms of hydrogen combined with atoms of oxygen, there was no water.

As explained in Part I, the protons and electrons created in the early stages of the Big Bang had to wait about 380,000 years to combine into ordinary hydrogen, and what's more, the creation of oxygen took place much later. Heavier atoms such as oxygen are formed inside stars, and the first stars began to form some hundred million years after the beginning, long after the epoch addressed in these first two verses of Genesis. In the time before stars, there was no oxygen. And without oxygen, there was no water. Even though all the translations above translate the Hebrew word in question as *water*, the idea that there was any water during this early epoch must be rejected. If there was no water during the time addressed in this passage, then why did Moses use the Hebrew word for water?

Well, apart from the Godly quotes, Moses had to select words that were part of his vocabulary to describe what he saw, and it's safe to assume that Moses did not know words such as proton, electron or plasma. So how did Moses come by this information? The one-word answer is *inspiration*. Volumes have been written on the topic of inspiration, but we have no scientific or biblical evidence to explain the particulars of what happens during an inspiration event. Inspiration is the handiwork of God, and God chooses whatever mechanism He wishes to accomplish His purposes.

In addition, we have no reason to believe that God acts in the same manner each time He provides inspiration. In Moses' case, it's clear that the impetus to write was part of the process, but what to write was a separate matter. The Godly quotes needed no explanation or interpretation. God said whatever God said and Moses wrote those words down, verbatim. Knowledge of what happened as a result of God's words had to be conveyed in some manner to Moses. Then Moses chose words that were part of his vocabulary to describe what he saw. Once again, what happened?

What did Moses see that caused him to choose "water" to describe it?

As previously discussed, the evidence from science indicates that the epoch described in the headline spanned about the first 380,000 years of the Big Bang, during which the cosmos was dark. In addition, my conjecture is God provided Moses a view of the cosmic expansion from the perspective of one who watched from the outside. Based on what he saw, "water" was the best word Moses could draw from his vocabulary to describe this deep, dark thing that was expanding as he watched. Viewing an ocean from above, at night, seems to fit what we imagine Moses saw; it looked like the surface (*the face*) of a body of water.

The late Stephen Hawking taught us that even black holes can emit a bit of radiation. Perhaps Moses saw a few photons of light scattering about at the surface of the expanding cosmos. In this speculation, the face he saw may have had a bit of shimmer as one might see looking down at a body of water at night. If the inspiration took place in this way, what word, other than water, might Moses have chosen to better describe what he saw? Regardless of exactly how God revealed the information to Moses, well established science shows that there was no water during the first 380,000 years of the cosmic expansion. Still, Moses did write a very impressive description of what one might expect to see from an outsider perspective, and his description corresponds with the scientific claim that the early cosmos was dark.

Thus far, our comparisons have been straightforward and the conclusions clear. In fact, the cosmos did have a *beginning*. In fact, the early Earth *was,* and in fact, it was *without form and void*. What's more, even though the young cosmos was made overwhelmingly of light, in fact, it was *dark* because the plasma was opaque. Even though science and scripture employ different terms to express these claims, the two sets of claims are synonymous.

However, solely on the basis of well established science, the translation choices of *wind* and *water* cannot stand scientific scrutiny. If wind is understood as movement of Earth's atmosphere over the surface of the planet, then there was no wind. Furthermore, there was no water. Instead, the *Spirit of God was hovering over the face of the deep,* hovering over the expanding hypersphere of plasma. Thus far, our thesis is upheld: science and scripture are telling us the same truths. With the above observations in mind, consider once more the genius of the first few words which form the headline for the rest of Genesis.

In the beginning God created the heavens and the earth. The earth was without form, and void; and darkness was on the face of the deep. And the Spirit of God was hovering over the face of the waters. (Genesis 1:1-2)

Truth Claims Following The Headline

Following the headline, the text continues with specifics. Careful reading shows that in some cases, God created something instantaneously, and in other cases, God initiated a process to bring about the desired results. For instance, in the text just below, when God said, "Let there be light," the light appeared instantaneously. This is what the church calls *creatio ex nihilo*, which means creation out of nothing.

Further along in the creation story, we read that God repeatedly commanded the beginning of processes that would achieve God's goals. For instance, when God said, "Let the Earth bring forth..." He initiated a lengthy process, not an instantaneous outcome. Science lumps the collection of these processes into the basket we call evolution. With these distinctions in mind, we continue with the creation narratives set forth in the NKJV of the Holy Bible.

Then God said, "Let there be light"; and there was light. And God saw the light, that it was good; and God divided the light from the darkness. God called the light Day, and the darkness He called Night. So the evening and the morning were the first day. (Genesis 1:3-5)

Notice that God spoke just four words in this passage. The rest of the verse is the commentary of Moses in which he describes what happened as a result of what God said. The quotes need no explanation; God said whatever God said, and Moses wrote God's words down verbatim.

We continue the comparisons taking the last claim first: The evening and the morning were the first day. Science has confirmed that the cosmos required about 13.8 billion years to reach its present form, not just six, 24-hour days. We know a really big object struck the Earth about 65,000,000 years ago, a collision that wiped out the dinosaurs. We know several species of hominids existed during the last few million years. Given this science, how are we to understand the six days of creation? Well, my conjecture is that the six days mentioned in the account of Genesis were six days of Moses' viewing, not six days of God's creating. Now, back to the beginning of the passage.

In this, the first moment of creation, God instantaneously produced the energy (the light) from which everything else would be formed. As discussed in Part I, during the early stages of the Big Bang, the photons of light outnumbered the other particles by a ratio of about 1,000,000,000 to 1. This means the cosmos was immersed in unimaginable light. So the first truth claim in this passage coincides with the truth claim of science, but science is understandably silent on the claim that God created the light.

Another truth claim in the passage above is that God divided the light from the darkness. This is an extraordinary truth claim! It's extraordinary because light and darkness do not coexist. If light is present, there is no darkness; if darkness is present, there is no light. For illustration, if one enters a room with no windows or other active source of light and closes the door, then the room is filled with darkness. If one switches on the light inside the room, then the darkness is gone. These two, light and darkness, do not exist in the same space at the same time. Given this science, how are we to understand the claim that God divided two things that do not coexist?

Again, as explained in Part I, for about the first 380,000 years from the beginning, the cosmos was dark because, even though the light was present in overwhelming abundance, it could not shine because the plasma of charged particles was opaque. The light was trapped inside. In this unique case, the light and the darkness did coexist; they did occupy the same space at the same time. However, when the temperature dropped sufficiently, the electrons entered orbits about the nuclei, and the cosmos became transparent.

At this moment, the light was separated from the darkness. Notice the timeline of these two events. From the first moment of creation, to the dividing of the light from the darkness, 380,000 years passed on the cosmic calendar. The above scripture coincides with what we have learned from science, and scripture gives God the credit for making it happen. What began with an instantaneous creation event continued with the processes of expansion and cooling. At the right time, the cosmos became transparent to the light, and this was the moment the light and dark were separated.

In verse 5 we read that God called the light Day and God called the darkness Night. Modern people understand one day to be the time required for the Earth to make one revolution about its axis. Each day takes 24 hours. When God called the light Day (notice the capital D), He was not speaking of the kind of day that depends on the rotation of the Earth. Although sometimes confused, these are different understandings of what Day actually is. During the era described in the preceding passage, there was no Sun and no Earth in their present forms. This means our present definition of one day could not have been applicable back then.

The next passage leaps forward billions of years from the beginning. Some folks have little idea of the magnitude of such numbers. Unless you happen to be a scientist (or politician) you probably don't often use the word "billion," let alone "trillion". For illustration, suppose a man who is six feet tall becomes a million times taller. If this fellow lies down (in the right direction) with his head near Little Rock, Arkansas, then his feet will be somewhere down in Texas. Now, suppose this fellow becomes a billion times taller. If he stands up, the moon would be somewhere around his knees. Back in his prone position, if we make this same guy a trillion times taller, his body will go completely around the earth over 45,000 times. A billion years is a long time.

By this point in the evolution of the cosmos, some of the first-generation stars died when they blew themselves apart in stupendous explosions called supernovae. The oxygen and other materials they had produced during their lives was dispersed and later incorporated into second-generation stars like our Sun. Some of the dusty materials didn't make it into the second-generation stars. They remained in obit around the stars and over time, formed planets. We live on one we call Earth. At this point in the evolution of the cosmos (about 9.3 billion years from the beginning) there was an earth with form, and there was water. Here is the text.

Then God said, "Let there be a firmament in the midst of the waters, and let it divide the waters from the waters." Thus God made the firmament, and divided the waters which were under the firmament from the waters which were above the firmament; and it was so. And God called the firmament Heaven. So the evening and the morning were the second day. (Genesis 1:6-8)

Unaware, as they surely were, that there were billions of other galaxies of stars, a number of which had planets and moons, early commentators took the two categories of water mentioned in this passage to mean only waters on the surface of the Earth and waters in the clouds just above. This view was not incorrect, just incomplete. Today, we know there are trillions of other stars, many with planets, and some of those planets have moons. It is a virtual certainty that liquid water exists in many locations other than Earth. As this is being written, reports of subterranean water on Mars are making news. Notice also, that the separation of the waters above from the waters below was not an instantaneous event. It takes time for water (or anything else) to move from one location to another. Depending upon the speed of water and the distance to be covered, the time required may have been small, or it may have taken thousands or even millions of years; either way, time was required. So without question, this passage speaks to a process, a pattern we will see repeated in the following passages. Science makes no claims regarding Heaven, and since we are comparing scripture with science, we move along to the next passage.

Then God said, "Let the waters under the heavens be gathered together into one place, and let the dry land appear"; and it was so. And God called the dry land Earth, and the gathering together of the waters He called Seas. And God saw that it was good. (Genesis 1:9-10)

Earth today looks very different from the way it looked in the distant past. The scientific evidence suggests early earth was covered by water. The details are debated, but I think it is safe to say most geoscientists believe a very large land mass was pushed up from below the surface of the water to become the super continent called Pangea. As Pangea rose to the surface, it shoved the primordial ocean aside and dry land appeared, just as the passage from Genesis above claims. Pangea subsequently broke apart into two other large land masses known as Gondwana and Laurasia.

These histories cover billions of years. We are certain the continents are continuing to move about over the surface of the planet today. We know South America and Africa were part of one landmass in the distant past, but today they are thousands of miles apart. This science fits very well with the truth claim in the above passage. Furthermore, this is another example which shows God often initiated processes in order to accomplish His purposes. Scientists call these geologic phenomena "plate tectonics".

As will be the case in all instances of God pronouncing something good, science is silent. Notice this passage does not end with the announcement of the end of the third day. It seems that Moses kept watching, listening and writing. This brings us to the next passage of interest.

Then God said, "Let the earth bring forth grass, the herb that yields seed, and the fruit tree that yields fruit according to its kind, whose seed is in itself, on the earth"; and it was so. And the earth brought forth grass, the herb that yields seed according to its kind, and the tree that yields fruit, whose seed is in itself, according to its kind. And God saw that it was good. So the evening and the morning were the third day. (Genesis 1:11-13)

In this passage God commissioned the Earth to bring forth three kinds of plant life: grass, herbs and fruit trees. This was the beginning of processes, not instantaneous creation events. We lump these processes into one basket which we call evolution. The plants paved the way for animal life that would come later. Plants (at least the green ones) harness the energy of the Sun through a process known as photosynthesis, and when animals consume the plants, they use this plant energy to carry on their own affairs. As is logical, there are no animals mentioned in the scripture before the era of plants. This passage includes the pronouncement of evening and morning as the third day; not the third day of God's creating, but the third day of Moses' viewing.

Then God said, "Let there be lights in the firmament of the heavens to divide the day from the night; and let them be for signs and seasons, and for days and years; and let them be for lights in the firmament of the heavens to give light on the earth"; and it was so. Then God made two great lights: the greater light to rule the day, and the lesser light to rule the night. He made the stars also. God set them in the firmament of the heavens to give light on the earth, and to rule over the day and over the night, and to divide the light from the darkness. And God saw that it was good. So the evening and the morning were the fourth day. (Genesis 1:14-19)

This passage makes truth claims about the formation of stars, and as discussed in Part I, the first stars began to form some tens of millions of years after the beginning. These first generation stars were formed of the hydrogen and helium created during the early stages of the Big Bang, and during their lifetimes, these stars fused the lighter elements into a series of heavier elements such as carbon, oxygen, nitrogen and iron. As mentioned earlier, a number of these first generation stars ended their lives in supernovae. These explosions seeded the surrounding space and other clouds of pristine hydrogen and helium with the heavier elements they produced before they exploded. These heavier elements are abundant in the dust of the Earth and in second generation stars such as our Sun. Here, we see another scriptural truth claim that corresponds with science: the star we call the Sun (the greater light in the passage above) was formed after the other lights (first generation stars) were formed. Skeptics may point out that since green plants require sunlight, this history cannot possibly be historically accurate for it places the arrival of plants (Genesis 1:11-13) before the creation of the Sun. Fair enough, but as we learned from Carl Woese, the three earliest forms of life were quite different from the green plants or the animals we see around us today.

Furthermore, when God said, "Let the earth bring forth...," He initiated a lengthy process, not an instantaneous outcome. The earliest forms of life did not require sunlight for their primary energy source. We have come to understand a good deal about these evolutionary processes, and nothing in the passage above contradicts the science.

Then God said, "Let the waters abound with an abundance of living creatures, and let birds fly above the earth across the face of the firmament of the heavens." So God created great sea creatures and every living thing that moves, with which the waters abounded, according to their kind, and every winged bird according to its kind. And God saw that it was good. And God blessed them, saying, "Be fruitful and multiply, and fill the waters in the seas, and let birds multiply on the earth." So the evening and the morning were the fifth day. (Genesis 1:20-23)a

The truth claim in this scripture is that living creatures first appeared in the waters. It has turned out that this claim corresponds well with what we have learned from science.

We have good evidence which shows North America and Europe were one mass of land in the distant past; likewise for South America and Africa. Obviously, these continents are no longer connected, and they continue to distance themselves. As a result, the Atlantic Ocean grows a bit wider each day. T

he place where the division originates is known as the Atlantic Rift. On one side of the rift, the sea floor is moving east, while on the other side of the rift, the ocean floor is moving west, and new sea floor is forming at the junction of spreading. This rift is a very active piece of watery real estate. It is here, in the active spreading region, that one class of possible candidates for a suitable location for early life has been found. Technically, they are thermal vents, but we call them black smokers.

Black smokers spew out jets of super heated water rich in various chemicals. They form when seawater seeps into cracks in the ocean floor and sinks down where it is heated by the hot rocks below. The water is heated to several hundred degrees Fahrenheit and forced back to the ocean floor, where is spews out its load of dissolved materials into the cold waters that surround the vents. These minerals often are black in color, hence, the name. They are very hot, which means they carry energy and they carry dissolved materials, both of which are necessary for life. Could these black smokers be the birth places of early forms of life? Possibly, but objections to this view have been raised.

The temperatures of these black smokers can reach several hundred degrees, a situation not particularly conducive to life. Nevertheless, there are exotic forms of life called *extremophiles* that live in these neighborhoods. Life requires energy, and if the black smokers are just too hot for some early life forms, where else may we find more favorable energy conditions?

Well, in other locations of the ocean floor we find another class of thermal vents, and these also carry materials and energy, but not at such extreme temperatures. In December of 2000 another class of thermal vent was discovered by researchers on the ship Atlantis. These vents also spew out jets of hot water, but these jets are not as hot as the black smokers. The ejected water carries thermal energy along with mixtures of minerals encountered and dissolved in transit. So where are these vents located?

Most of the ocean floor is composed of basalt, a black, fine-grained, volcanic rock. However, there are large underwater mountains that are made of a very dense, hard rock called *peridotite,* which is usually found much deeper within the earth. Peridotite is green in color, and these mountains of peridotite are thought to have been brought to the surface and lifted up to form what is called the Atlantis Massif.[35] Massif means a very large mass of rock, and the first part of the name comes from the research vessel whose crew discovered them. The Atlantis Massif is approximately the size of Mt. Rainier in Washington State. It's about 14,000 feet high and about 10 miles across, and its peak is only about a half-mile beneath the surface of the ocean. Large fields of thermal vents populate the Atlantis Massif, so many that they are sometimes referred to as the Lost City.[36]

As mentioned above, these vents are not as hot as black smokers, but they are energetic, and they do eject abundant materials that early forms of life could have used. Maybe the Lost City thermal vents were havens for early forms of life. Or maybe life really did arise in Darwin's warm little pond. We cannot establish with certainty where the first life arose, but a watery environment was the birthing room, just as the passage of scripture claims.

Whether the first kinds of life originated in Darwin's warm little pond or near thermal vents on the ocean floor doesn't really matter for our purposes. What does matter is the realization that God commissioned the waters to the task. Regarding these historical events, science and scripture are complementary, not contradictory. God blessed these living creatures and instructed them to be fruitful and fill the water. Birds were also to multiply. This segment of history spanned a long time, and it seems to have taken Moses another day to watch it unfold. Moses concluded this section in his usual pattern; *the evening and the morning were the fifth day.*

Then God said, "Let the earth bring forth the living creature according to its kind: cattle and creeping thing and beast of the earth, each according to its kind"; and it was so. And God made the beast of the earth according to its kind, cattle according to its kind, and everything that creeps on the earth according to its kind. And God saw that it was good. (Genesis 1:24-25)

This is the account of God calling forth the third primary kind of life called living creatures. These would become land-dwelling animals in contrast to earth-based plants and the other animals which arose in the waters.

In 1977, Carl Woeses overturned one of the major dogmas of biology.[37] Until that time, biologists had taken for granted that all life on Earth belonged to one of two primary lineages: the eukaryotes (which include animals, plants, fungi and certain unicellular organisms such as paramecium) and the prokaryotes (all remaining microscopic organisms). Woeses discovered there were actually three primary lineages. Within what had previously been called prokaryotes, there exists two distinct groups of organisms no more related to one another than they were to eukaryotes. Because of Woeses' work, it is now widely agreed that there are three primary divisions of living systems: Eukarya, Bacteria and Archaea.

The new group of organisms (Archaea) was initially thought to exist only in extreme environments, niches devoid of oxygen and whose temperatures can be near or above the normal boiling point of water. Microbiologists later realized that Archaea are a large and diverse group of organisms widely distributed in nature and common in much less extreme habitats, such as soils and oceans. In line with our thesis, science claims there were three different kinds of ancient life, and so does scripture.

God called forth three kinds of ancient life through processes, not as instantaneous events. Unless one is prepared to believe giant sequoias and stately cedars shot up out of the soil as though propelled upward by a subterranean elevator, and to believe whales, dolphins and the other denizens of the deep instantaneously filled the waters as fully formed creatures and to believe, even further, that fully feathered eagles, hummingbirds and peacocks popped out of the surface of the water like popcorn from a kettle, then reason demands we embrace the idea that God initiated processes which gradually bought forth different kinds of life. The scientific evidence is very clear on this point. It may be no more than coincidence, but as Carl Woese taught us, the early Earth was home to three primitive kinds of life. So, other than the claim that God caused these three kinds of life to come forth, science and scripture are telling us the same stories.

Notice that this passage does not conclude with the declaration of the evening and the morning as another day. This passage, plus the story of the males and females in verses 1:26-31 just below, form a unit. This unit describes Moses' sixth day of viewing.

Job descriptions, including the blessings and locations are important in our quest to properly understand the texts, but one particular issue complicates our task. How are we to understand what it means to be created in the image of God?

Then God said, "Let Us make man in Our image, according to Our likeness; let them have dominion over the fish of the sea, over the birds of the air, and over the cattle, over all the earth and over every creeping thing that creeps on the earth." So God created man in His own image; in the image of God He created him; male and female He created them. Then God blessed them, and God said to them, "Be fruitful and multiply; fill the earth and subdue it; have dominion over the fish of the sea, over the birds of the air, and over every living thing that moves on the earth." And God said, "See, I have given you every herb that yields seed which is on the face of all the earth, and every tree whose fruit yields seed; to you it shall be for food. Also, to every beast of the earth, to every bird of the air, and to everything that creeps on the earth, in which there is life, I have given every green herb for food"; and it was so. Then God saw everything that He had made, and indeed it was very good. So the evening and the morning were the sixth day. (Genesis 1:26-31)

The difficulty we encounter in Genesis 1:26-31 is this passage has traditionally been understood to refer to the creation of Adam and Eve, since these folks were created *in the image of God*. Even though this interpretation is part of the deep ruts mentioned earlier, this passage does *not* refer to Adam and Eve. God created Adam in what we would call a hands on, one-off manner, and sometime later God cloned Eve from Adam's rib. The passage above describes a process that God initiated to bring about the desired results. This process would culminate in males and females with the capacity to exercise dominion over the other animals.

God has absolute dominion over everything, and to be created in the image of God seems to suggest that those who were blessed received a measure of this quality which God possesses without limit. We are certain that several other species of hominids lived on this planet over the past few million years. I am entertained by the idea that God intervened at some point along the path of evolution and blessed one species with added abilities. The blessing could have come as a change in their DNA. We know anatomically modern Homo sapiens show up in the fossil record inside Africa at about 300,000 years ago. As discussed in Part II, these hominids transitioned into fully modern Homo sapiens in the relatively recent past, and since these folks are still around and all the others are gone, they (we) surely were the ones with dominion.

God lit the fires of evolution that led to various forms of life, rather like a spark lights a pile of kindling which grows into a roaring fire, but at least for the most part, it seems that God let the fires of evolution burn without divine intervention. Notwithstanding, the passage which claims that God created these folks *in the image of God*, indicates that, at least in this case, God did intervene in the evolution of hominids. Lacking any evidence to the contrary, my conjecture is to be created *in the image of God* is indicative of being given dominion over the other forms of life. God has dominion over all things, and these creatures were given a measure of this Godly ability.

What other evidence is there to substantiate or falsify the claim that Genesis 1:26-29 is not about Adam? Keeping in mind what we have learned from science, we compare scripture with scripture.

First, notice that the males and females mentioned in Genesis 1:26-29 did not speak, or if they did, there is no record of it. To assume they could speak is to let the Sitz im Leben hold sway over what the text actually says. These creatures heard God speak, but so did the water and the earth. Like the water and earth, these males and females made no verbal response.

Second, God told them to be fruitful and multiply, to fill the earth and subdue it, and to have dominion over the other forms of life. Obviously, it would have taken a rather long time for these folks to have enough children to fill the earth — a long process, not an instantaneous outcome.

Third, God gave them a menu that included herbs and trees wherever they found them, not only in the Garden of Eden, but over the entire planet. Obviously, these folks came along after — not before — God called forth plant life from the earth. This will be an important consideration when we come to Genesis 2:4-5. The story of the males and females in the first chapter of Genesis is very different from the story of Adam and Eve which begins in the second chapter of Genesis. The text is clear that Adam and Eve did speak, and there are other differences between the two lines of hominids.

Fourth, Adam was as a bachelor, and bachelors (at least those without sexual partners) do not have enough children to populate a small garden, let alone the entire planet. Quite naturally, God did not tell Adam to multiply and fill the Earth. Why would He have? God had already assigned this employment to the males and females mentioned earlier. Adam's bachelorhood was however, not permanent. But before God cloned Eve from Adam's side, God brought the animals to see what Adam would name them. We have no evidence, either way, but assuming there were about as many different kinds of animals back then as there are today, and assuming further that it took a reasonable period of time for the animals to gather for this event, this naming episode took Adam more than a few hours. We can say this episode took place in the relatively recent past, which means that velociraptor and T-rex were not among the attendees. We are not given a list of the names of the animals, for this is not the point. We are only told that whatever Adam called them, that was their name. The point is Adam was a symbolic thinker. He had the cognitive ability to realize that objects (animals in this case) can have names, and he was capable of complex language which he employed in naming them. Adam was a bachelor who didn't have children for some time, but he had neighbors.

Fifth, and in contrast to those God told to fill the Earth, God put Adam in a specific location where God had previously caused various kinds of trees to grow. We know this special place as the Garden of Eden.

Sixth, Adam's vocation in the garden was to tend and keep it. This means Adam did till the soil.

Seventh, concerning the menu, God told Adam that he could eat from every tree in the Garden except *the tree of the knowledge of good and evil.*

Adam was a fully modern Homo sapiens. He realized that objects can have names. He was capable of symbolic thinking and complex communication, and he had the cognitive wherewithal to engage in agriculture. Actually, Adam was superior to modern people, at least in some ways.

Adam lived a long time by today's standards. No one lives 200 years in these days, let alone 933 as did Adam. As pointed out in Part II, the evidence shows farming was underway by roughly 12,000 years ago, not in Africa, but in the Fertile Crescent, which by the way, was the geographic location of the Garden of Eden, the home of Adam.

So, the story of the males and females in the first chapter of Genesis is very different from the story of Adam and Eve that begins in the second chapter. Attempts to forge the two stories into one have only made things worse. The genealogy in Genesis 2:4-6 is not about Adam. The males and females who were given dominion over the other forms of life were not Adam and Eve.

There were substantial differences between Adam and the other hominids who meandered through the Fertile Crescent from time to time, but one difference overshadowed all others. God gave Adam a soul.

Concerning The Soul

I want to know one thing, the way to heaven: how to land safe on that happy shore. God Himself has condescended to teach the way. He hath written it down in a book. O, give me that book! At any price, give me the book of God! I have it: here is knowledge enough for me. Let me be a man of one book.

— The Reverend John Wesley

Rev. Wesley was a well-educated English clergyman, fluent in Latin, German, Hebrew and Greek. Despite what he wrote in the above aphorism, he was a man of many books, a substantial number of which he authored. And for all his learning, he believed the most important thing in this life is making sure we "land safe on that happy shore" in the next. Wesley was convinced that God teaches the way to heaven — in the Bible.

When this life comes to an end, the soul will return to God in heaven, or it will be destroyed in hell. There are no alternatives. One or the other of these two will surely come to pass. Not only does the Bible teach the way to heaven, it also teaches us how to live in proper relationship with God and with other people.

The Hebrew word frequently translated as soul is נֶפֶשׁ (pronounced **neh**-fesh). The Enhanced Strong's Lexicon reports the following: *Nehfesh* occurs 753 times in the King James Version of the Old Testament, and it is translated as "soul" 475 times "life" 117 times, "person" 29 times, "heart" 15 times, and there are other, less common translation choices. Brown Driver Briggs, a highly respected Hebrew and English lexicon, explains *nehfesh* as follows: נֶפֶשׁ famine noun translated as: soul, living being, life, self, person, desire, appetite, emotion and passion — 1. = that which breathes, the breathing substance or being; the soul, the inner being of man.[38]

The first instance of *nehfesh* in the Old Testament is in Genesis 2:7 where we read that God breathed the breath of life into Adam whereupon Adam became a living soul. (KJV) Other versions translate *nehfesh* as *being*. Being and soul are not equivalent. We'll come back to this crucial point a bit further down. For now, let it be enough to point out that every person has an appointment with the Grim Reaper, and when that day arrives, the only thing that will matter is whether their soul goes to heaven or to hell, for these are the only options.

If the soul does land in heaven, then it will live on forever. Otherwise, it will be destroyed. Yes, destroyed. Not tormented for all eternity. Although the idea that lost souls are tormented forever is deeply ingrained church dogma, multiple passages of scripture suggest a different view. Furthermore, the idea that souls which arrive in hell live on forever in torment is antithetical to the idea of a loving God. And it is contrary to the teaching of Jesus who said, *"And do not fear those who kill the body but cannot kill the soul. But rather fear Him who is able to destroy both soul and body in hell."*[39] Jesus also said, *"Enter by the narrow gate; for wide is the gate and broad is the way that leads to destruction, and there are many who go in by it."*[40] Furthermore, in what is perhaps the most famous passage in the New Testament, Jesus said, *"For God so loved the world that he gave his only Son, so that everyone who believes in him may not perish but may have eternal life."*[41]

Obviously, the eternal life Jesus mentioned has nothing to do with the physical body, for the body of every person will certainly perish, no matter what they believe. It's the soul that has the potential to live on after the body dies. As concerns the immortality of the soul, the words of interest in the three passages just above are: *kill, destroy, destruction* and *perish*. Each of these four words denote a terminal event, which means that souls which arrive in hell are obliterated. There is nothing in these passages to suggest eternal torment of a soul. Yes, hell is the place of eternal fire, but it was prepared for the devil and his angels.[42] Souls do not live on in that place. If human beings do have a soul, what could be more important than making sure it returns to God in heaven?

If we live a hundred years or more during which we attain all our dreams and fulfill all our ambitions, but do not arrive in heaven when we die, then in the final analysis, we are miserable failures. Nothing is more important than our soul landing safe on that happy shore, absolutely nothing. Earlier, we cited two scholarly sources to gain insight into the Hebrew word *nehfesh*. How else might we think about the soul?

We can say a soul is a unique collection of information that fully and precisely describes the life of one person. Even though certain bits of information will be commonplace, no two souls are identical. This is rather like the DNA of all people being quite similar and yet, unique. What else can we say about a soul?

This Call May Be Recorded

The soul is reminiscent of that certain phone call that many have made, the one that begins with the answering party announcing, "This call may be recorded." In this manner, the caller is made aware that a record of his or her words may (and almost certainly will) be made, and obviously, the record of this conversation will be stored somewhere. Although we may not be able to establish with precision where something is, if it exists, it must be somewhere. Regardless of where any information is stored, if it is to have value, it must also be retrievable. The phone recording, per se, has nothing to do with any idea of good and evil. It only records what is said. Those who might refer back to the recording can decide the merits of the information for themselves. Likewise for souls, they only contain the data, and God will judge the merits of the information. Exactly how the soul stores the information, we cannot say with certainty, but we can say, "This call is being recorded."

The soul goes further than the phone call. In this metaphor, it records not only a single conversation, but all of the words we've ever spoken. It also records all our thoughts, our emotions and our actions. In short, the entirety of our life is being recorded in the package we call the soul. Even though we would like to know the soul's exact location and precisely how it works, for the present, we will have to settle for knowing that, wherever it is located and however it works, it stores the information without error. That the soul performs its function without error is not surprising because souls come from God and belong to God. [43]

Through science we have learned that there are two main systems within the human body that store and process information, the brain and DNA. Obviously, the information contained in the brain is centralized inside the cranium, but the information contained in DNA is duplicated trillions of times inside the cells. Over the last century, we have learned a good deal about how each of the two systems do the things they do, but we cannot yet fully describe the operation of either.

We know the mature brain processes information, and we are presently convinced that memory exists within the billions of cells located inside the skull. Interestingly, the location, size, configuration and operation of the brain depends on DNA, for it's the DNA that determines whether a particular cell becomes part of the brain, or part of a bone, or part of something else in the body. The brain does not direct the construction of DNA. To the contrary, DNA directs the construction of the brain. Presumably, the brain contains relatively little information at birth. Obviously, the content increases as the years go by and experiences accumulate. In contrast, DNA comes preloaded with a huge amount of information, including the information required to build a brain and all the rest of the body, plus the information required to keep it functioning.

We know DNA is a very complex information storage and retrieval system, and I am highly entertained by the idea that DNA holds the information we call the soul. We know a zygote (a fertilized ovum) does not have a brain, so either these earliest forms of human life have no souls, or if they do, their soul resides somewhere other than in the brain. If souls are somehow part of the DNA, then each person receives a soul at the moment of conception.

Do we have any evidence to support the conjecture that DNA houses the soul? We might.

Colton Burpo was three years old when he underwent surgery for a ruptured appendix. Following his recovery, he reported to his parents that he had gone to heaven during the surgery. Colton had an older sister at the time, but prior to his surgery, he did not know that his mother had a miscarriage before he was born. In describing his heavenly experience, Colton told his mother that he had a sister who had "died in her tummy" before he was born. In addition, he told his mother he met this previously unknown sister in heaven. Colton's account lends credence to the idea that unborn children do have souls. His sister's soul landed safe on that happy shore. In this view, the soul is not located inside the brain. Instead, it is located within the DNA. Colton's book is entitled *Heaven is for Real*.

British physicist Sir Roger Penrose and Professor Emeritus Stuart Hameroff, M.D., have suggested that quantum vibrations within microtubules in the brain's neurons interfere, collapse and resonate — that they control neuronal firings, generate consciousness and ultimately connect to ripples in spacetime geometry. They call these phenomena "orchestrated objective reduction" or "Orch Or" for short.[44] Speaking as they do of quantum phenomenon, they propose heady stuff, even for those with a bit of scientific training. Not surprisingly, their proposal has been criticized, especially by atheists. If I have properly understood what they mean, they seem to have at least some idea of how we might begin to understand consciousness, and by implication, the soul.

Atheists would deny that human beings have a soul, at least in the Christian sense of the word. Even so, they could not deny that memory is a real phenomenon. Ask an atheist their name or birthday. If they answer, they will have to admit they retrieved the information from memory. Atheists accept the idea of memory as stored information, just not the kind of stored information Christians call the soul. Obviously, we either have a soul or we do not.

Given that souls are real, then does everything in the cosmos have one? Scripture suggests only living things have a soul. Inanimate things such as water and rocks do not. Only living organisms have souls and only one kind of organism out of the many. As discussed earlier, Carl Woese taught us there were three primordial kinds of life, and we doubt that anyone would claim these earliest forms of life had souls. Neither bacteria, nor viruses, nor plants, nor any other lower forms of life have souls. Chimpanzees, gorillas, lemurs and dogs have DNA, but they do not have souls.

Charles Darwin taught us that different kinds of life evolved into many other — and generally more complex — kinds of life. However, souls did not come into existence through evolution. Still, at some point during the eons of evolution, creatures who do have souls came onto the stage of history. For the most part, science is silent on matters of the soul. In our efforts to better understand souls and the beings who have them, we turn to scripture.

The First Soul

As noted earlier, the first mention of soul in the Bible is found in Genesis 2:7 where we read the account of God creating Adam. The biblical texts that precede this one-off event make no mention of a soul. So unless our Sitz im Leben drives us to read more into these preceding texts than they actually contain, Adam was the first man to have a soul. Since God created Adam in a hands-on manner, Adam certainly had perfect DNA, for the idea that God put faulty DNA in Adam is simply beyond the pale. Adam's perfect DNA gave him superior cognitive abilities, a life span over nine centuries and free will. God gave Adam a soul. Since Adam was the first hominid to have a soul, Eve was the second because she was cloned from Adam's body. Her DNA was almost identical with Adam's. The only difference was God gave Eve two copies of the X chromosome which made Eve female. One X and one Y makes the individual a male. If DNA really does contain the soul, then the descendants of the Garden couple also had a soul, and that from the moment of conception.

Not everyone agrees God made Adam as a living soul. The NRSV, NIV and NKJV Bibles translates *nehfesh* as "being." In their view, Adam became a living being. For convenience, just below are the KJV, NKJV, NIV and NRSV translations of Adam's creation.

And the LORD God formed man of the dust of the ground, and breathed into his nostrils the breath of life; and man became a living soul. (KJV)

... then the Lord God formed man from the dust of the ground, and breathed into his nostrils the breath of life; and the man became a living being. (NKJV)

And the Lord God formed man of the dust of the ground, and breathed into his nostrils the breath of life; and man became a living being. (NIV)

... the Lord God formed the man from the dust of the ground and breathed into his nostrils the breath of life, and the man became a living being. (NRSV)

Each of these four translations agree that God created Adam from the dust of the Earth. They also agree that God breathed the breath of life into Adam's nostrils. Furthermore, they agree that God's actions brought the inanimate dust to life. However, they differ on how Adam is to be understood.

The KJV claims Adam became a living *soul,* but the other three claim that Adam became a living *being.* Obviously, Adam was alive. But was he just one more living being among the many others, or was he a living soul? This is not a trivial question.

The word "being" denotes something which is alive; there are no dead beings. All living beings eventually die, and when one does, that being ceases to exist. What remains is a corpse, and the atoms of that corpse will return to the dust from whence they came. Even though a living soul resides within a living being, when the being dies, the soul does not. The soul departs, as happened to Rachel, the wife of Jacob. She died in childbirth, as explained in Genesis 35. Her being died, but her soul lived on.

When we say a person has a soul, we are saying the person is more than only their physical being. That which is physical will surely die, but the soul has the potential to live on. The destiny of the soul, whether it lives forever in heaven, or is destroyed in hell, is determined by a decision each person will inevitably make, whether they purposely choose to do so or not. Failure or refusal to make the right decision is equivalent to making the wrong decision. The information we need to make the right decision is set forth in the Bible, just as Rev. John Wesley believed.

Adam was a living being, as were the other hominids who came onto the stage of history through what we call evolution. However, Adam was more than only a living being. This is interesting. Surely, the physical body is part of a person's being, and when the person dies, the physical part of being comes to its end. This begs the question of what a being is, if it is dead. Stated differently, what would "dead being" mean? Well, it's nonsense; there are no dead beings. The King James Version is correct: Adam became a living soul, and what's more, he became the first person to have a soul.

The imperfect biological memory system we use every day sometimes fails us when we need to recall names, appointments, birthdays or anniversaries and such, but the soul is a perfect record of everything a person has thought, felt, said and done throughout their life. Because it is perfectly accurate, the soul contains those episodes we remember with pride, as well as those we remember with shame. In general, the shameful stuff is the result of sin, and whether we care to admit it or not, we each have some of both. The Apostle Paul stated it this way, "*... all have sinned and fall short of the glory of God.*"[45] Our memory may forget some of our sins, but our soul never forgets any of them.

By way of analogy, the soul holds information as does a computer hard drive, but there are important differences between the two. Hard drives operate at the pleasure of the computer user, and the drives can be disabled or turned off when not needed. Undesirable information on a computer can be edited or erased altogether, but information in a soul cannot. We have heard stories of people trying to erase seedy emails or lurid photos from their computers. However, the information in the soul, no matter how shameful, cannot be edited or deleted by a human being. Even though the soul is an integral part of each person, and even though human beings write the data to their soul, and even though the soul is the most important part of the person, the person is not the owner of the soul. Souls belong to God, and not only do they function flawlessly, they return to the Owner when a person dies. If the soul pleases God, it will live forever in heaven.

We've said a good deal about the soul because the final state of the soul is so very, very important. For emphasis, here again is Jesus' assessment of the issue:

"For what profit is it to a man if he gains the whole world, and loses his soul?" Matthew 16:26

Anthropic Principle

When I consider Your heavens, the work of Your fingers, the moon and the stars, which You have ordained, what is man that You are mindful of him, and the son of man that You visit him?

— A Psalm of King David

In comparison to the heavens, mankind is a tiny and seemingly inconsequential part of the whole, and yet, God is mindful of man. Not only has God created a hospitable environment for us, He has gone to great lengths to ensure that things end well for His frail creatures. The Apostle Peter wrote, *"The Lord is not slack concerning His promise, as some count slackness, but is longsuffering toward us, not willing that any should perish but that all should come to repentance."*[46]

God does not want anyone to perish. The idea that God is just watching and waiting for us to sin, whereupon He will throw us into the fires of hell where our soul will be tortured for eternity may turn out to be the most spectacular misunderstanding of God. God loves us, and God has taken extreme measures to ensure things end well for those who love and trust Him. Psalm 8 cited above is difficult to reconcile with the eternal punishment scenario.

It is a scientific fact that the universe is finely tuned in such a way that human beings can be part of the whole. As explained in Part I, there are a number of physical constants that determine the nature of the cosmos, and if these constants were slightly different from the values we measure them to be, then the cosmos would not exist in its present form.

Furthermore, whatever form it might take, human beings would not be around to ponder any of it. For example, if gravity were a bit weaker, then the early cosmos would have flown apart before complex structures could form, in which case we would not be here. If gravity were slightly stronger, then the cosmos would have collapsed back upon itself, in which case, there would be no stars, planets, moons or human beings. There are dozens of these constants, and the fact that these constants are finely tuned to values they must have in order for human beings to exist is called the anthropic principle (from the Greek, *anthropos*, which means human being). This fine tuning is a thorn in the side of scientists who are atheists because we have no scientific theory to explain why these physical constants actually do have the values they must have in order to have a cosmos friendly to human beings. Although not all Christians are aware of the fine tuning of these physical constants, those who are would say God tuned them for us. In this view, we might catch a glimpse of God's fine tuning activities in the headline of Genesis.

We know that in order for the cosmos to evolve in the manner it has, the fine tuning had to take place at (or very soon after) the beginning, and although we have no idea how God did these things, perhaps we can get some idea of what it looked like from Moses' perspective. As mentioned above, we read in Genesis 1:2c, *And the Spirit of God was hovering over the face of the waters.* We also pointed out that the Hebrew word we translate as *hovering* is a word one would use to denote the action of a bird hovering over its chicks. So the episode described in 2c was not a quick fly-by. It was a more substantive encounter between the *Spirit of God* and the expanding hypersphere of plasma that was the young cosmos. My conjecture is that during this early encounter, God was tuning the constants to the values they must have in order for us to be here.

Knowing that the Bible does not mince words, and taking scripture literally and seriously, we are pointing out that the Spirit of God was hovering over the face of the deep at this time in the evolution of the cosmos, and although we cannot be sure, it is fair to say that God was doing something creative during the hovering episode. Interestingly, this is the only passage in the first chapter of Genesis that directly mentions the whereabouts of God. Christians who are familiar with the fine tuning of the constants would agree God tuned them, and He did so because He loves us. Human beings are a tiny, tiny fraction of the whole cosmos, but God loves us. God is *mindful of man* as the Psalmist wrote.

Multiverse

Knowing these constants must have the values they do in order for *Homo sapiens* to exist, and lacking any science to explain why they do, leaves the door open to the idea that God tuned them. For obvious reasons, atheists strongly dislike this idea, and rather than glorifying God for creating a cosmos that is friendly to human beings, they put forth a speculation called the "multiverse." People supporting this view are called, M-theorists. They speculate that our universe is one among many, possibly an infinite number of others. Specialists know that science doesn't do very well in scenarios that include infinities. What's more, one might rightly question the use of the word "theory" since there is no way to test their idea against evidence. However, for convenience we will use their terminology. According to M-theorists, since there are a huge number of other universes with physical constants that are different from the ones in our universe, we should not be surprised to find ourselves living in one with the physical constants whose values support life.

I recently asked a Nobel laureate physicist if he could imagine any way to determine whether or not our cosmos was infinite. He told me he could not. How then shall we talk sensibly about an infinite number of other universes outside our own? It seems, at least to me, that M-theory is no more than wishful thinking meant to obfuscate the handiwork of God in this universe, which by the way, is the only universe for which we have any evidence.

Theophany & Inspiration

A theophany is an encounter between God and an individual or group. In order for Moses to have been able to accurately describe the things he did, God provided the information Moses could not have discovered by any other means. Such a conveyance of information is called "inspiration." Even though our task is to compare and contrast the truth claims found in Genesis 1:1 – 6:7 with the corresponding truth claims of science, we temporarily step outside Genesis in order to consider theophanies and inspiration.

In the third chapter of Exodus, we read that Moses climbed Mount Horeb where the Angel of the Lord appeared to him as a burning bush. During this theophany, God gave Moses the ability to perform two kinds of miracles, along with the instructions on how to use them. The miracles were meant to coerce the Pharaoh of Egypt to release the Hebrew slaves. Moses demonstrated one miracle but not the other. Most have heard of Moses' staff turning into a serpent, a miracle that didn't much impress Pharaoh. The reader is left to wonder why Moses did not put his hand into his tunic to perform the leprosy miracle that God made available to him. The multiple confrontations between Pharaoh and Moses are recounted in the Book of Exodus. Even so, the several confrontations between Moses and Pharaoh concluded with Pharaoh releasing the slaves as recorded in Exodus 13.

Much of the Book of Exodus describes Moses receiving instructions from God on various topics, including the famous Ten Commandments. Multiple texts recount other instances where Moses received inspiration from God when Moses was not on the mountain. So from these we learn inspiration can be the result of a theophany in the form of a burning bush on a mountain, or as a word from God spoken into one's ear or mind.

Science is silent on the topic of Godly inspiration, and understandably so. Scientific theories make predictions that can be tested against the evidence, and these tests can be replicated on demand. However, God alone initiates and controls inspiration. Obviously, no one can put inspiration under the microscope for examination, and no one can falsify another's claim to have had an inspiration experience. Notwithstanding, anyone who believes that God has inspired (told) them to do anything which is contrary to scripture cannot be taken seriously. For example, if a person believes, even very sincerely believes, that God has told them to kill everyone with some unusual physical characteristic, they are very much mistaken.

Unfortunately, history is stained by the actions of those who believe they have had some strange word of inspiration from God. However, this does not mean there is no such thing as authentic inspiration. For illustration, one might be thoroughly convinced he owns a luxury yacht located on the Mediterranean Sea when, in fact, he does not. This person's delusion should not be construed to mean there are no luxury yachts on the Mediterranean. Hearing from God is these days is taken as problematic in some circles. In fact, Jesus said His disciples will hear His voice and follow Him.[47] Those who have not had a theophany usually doubt that others have, but for those who have had such an experience, no amount of rhetoric from those who have not can convince them otherwise.

It is clear from the accuracy of what Moses wrote that he was inspired to do so, for he described things that took place billions of years before he, or anyone else, was born. Moses came to know the unknowable. We cannot be certain, but whatever form the inspiration took, the events Moses described spanned billions of years, and his descriptions closely correspond to the scientific views of today. We could think of inspiration as a unique educational experience with God as the teacher.

Amidst the COVID-19 pandemic, many students participated in remote learning characterized by watching and listening to instruction on some sort of electronic device. We might liken Moses' inspiration to his watching a heavenly hologram or divine DVD. In this metaphor, the audio component of the hologram allowed Moses to hear what God said, and the video component gave Moses a view of what happened in response to God's words. Could God have supernaturally and instantaneously put the information into Moses' mind? We suppose so, but if the inspiration was instantaneous, then why did Moses spend 40 days on the mountain? We have no scientific or scriptural evidence to confirm or refute instantaneous learning, but quite frankly, this author's Sitz im Leben is such that a more gradual process seems more plausible. Moses was fully human, and part of what it means to be human is that learning takes time. What's more, Moses needed time to write down what he learned.

In addition, and although we might not often think about it, being fully human also means, at least under ordinary circumstances, that Moses got thirsty, hungry and tired. Moses also needed time to attend to those other daily tasks common to all people.

The phrase "ordinary theophany" is an oxymoron, for there is nothing ordinary about a human encounter with God. We can barely imagine how far from ordinary the theophany on Mount Horeb was.[48] As Moses stood barefoot on Holy Ground, in the presence of God manifested as a burning bush, perhaps the awe of being in God's presence interrupted some of the daily operations of Moses' body. Maybe God set aside Moses' ordinary bodily needs during this episode, but this is somewhat inconsistent considering that further along in the story of Israel, God provided the newly freed slaves from Egypt with water from a rock and manna from heaven as they journeyed through the desert.

Even so, just because one is hungry does not necessarily mean one eats right away. But thirst can only be denied a few days without severe dehydration. Without water, human beings die in just a few days. Although the passage from Exodus noted just above reports that Moses did not eat or drink during his time on the mountain, it does not mention whether or not Moses took intermissions to tend to other necessities. This is speculation. If

Moses did need intermissions, then obviously, the inspiration, which we have characterized as an educational experience, was interrupted from time to time. In this view, Moses watched and listened until he was overtaken by the need for an intermission, after which he again took up the task before him. Regardless of exactly how the inspiration came about, or how long it took to complete, it is clear from what he wrote that Moses learned how the cosmos evolved. This is evident in the fact that the truth claims of Genesis are synonymous with the truth claims of science.

All learning takes time, and the more material there is to learn, the more time it takes to learn it and to write it down. For example, we expect a high school education to take about 12 years; an undergraduate university degree to take an additional four years; a seminary degree, four more years, and so on. Learning, whether natural or inspired, takes time, and this is not surprising, for every change in the cosmos requires the passage of time. Changes within the nucleus of an atom may take only a very small fraction of a second. The change from a dark cosmos to one filled with shining light took about 380,000 years. The formation of the star we call the Sun took some hundred million years to transition from an enriched, swirling glob of cold, dusty, interstellar gas into a sphere hot enough to ignite the fire of nuclear fusion in its core. Even "instant" coffee takes a few moments to brew. The point is, change takes time. Inspiration, understood as the change from a state of ignorance to one of knowing, also takes time.

What's more, if the information exchange is wide-ranging, then it will take more time than will a short session on a single topic. For our purposes, we can say that the inspiration of Moses was the conveyance of information which Moses could not possibly have known in any other way. It is blindingly obvious that Moses was not around when God spoke the words that created the cosmos out of nothing, which, as mentioned earlier, is a longstanding church dogma known as *creatio ex nihilo*. Because the events Moses described spanned virtually the entire history of the cosmos, it is likely his learning session would have taken more than a few hours. These observations will be important as we come to the idea of a six-day creation.

Six Days of Creation

Continuing the metaphor of a heavenly hologram, it would have taken more than a few minutes for Moses to view the entire presentation which spanned 13.8 billion years of history. He had to assimilate what he saw and write it down. Perhaps it took about a week. In this view, we might catch a glimpse of why Moses wrote what we have come to see as the six days of creation. As mentioned earlier, my conjecture is that the six days in question were six days of Moses' viewing, not six days of God's creating. My best guess is the inspirational transaction, highly compressed as it was, took six days to complete, and these six days included intermissions, at least for Moses to write what he had heard and seen.

In the time of Moses, a *day* was understood to begin at sundown, to continue throughout the night and throughout the following hours of daylight. This day concluded and another began with the next sundown — *evening and morning.* This full cycle of darkness and light was equivalent to 24 hours, which we now understand to be the time required for the earth to make one revolution about its axis. Obviously, this 24-hour day (per one revolution of the earth) continues to be the modern understanding of one day.

Those who cling to the idea that God created the cosmos in six, 24-hour days, might want to consider what a "day" would mean in the saga before the earth and Sun were formed. From a scientific perspective, prior to the formation of the Sun and the earth, the idea of one day would be undefined.

In my speculation, Moses watched the unfolding of the cosmos until God paused the hologram. The next "day" the learning continued.

A New Day

It is understood that one day is the time required for earth to complete one revolution on its axis. However, there are two primary ways of designating this 24 hour interval of time. Civilians generally divide one day into two segments of twelve hours each and distinguish between the two by adding the a.m. or p.m. suffix to the number. For instance, it could be 12:00 a.m. or 12:00 p.m. The U.S. military uses a different system, one based on a 24 hour day without the a.m. or p.m. suffixes. Either way, and somewhat oddly, a new day on the clock begins in the middle of the night — 12:00 a.m. or 0000.

From a less formal perspective, a day begins with sunrise and ends with sunset. As a reminder, we have learned from science that the first epoch of cosmic history was characterized by unfathomable light, but as overwhelming as the light was, the cosmos was dark because the light was trapped inside the expanding hypersphere of plasma. The plasma was opaque, so the surface of the young cosmos was dark.

In light of this science, perhaps we can begin to understand why Moses reversed what might be called the usual order of things when he wrote, *the evening and the morning were the first day*, instead of writing the morning and the evening were the first day. He continued the pattern of evening followed by morning throughout this part of Genesis. Stated differently, since the first thing he saw was the dark cosmos, perhaps it makes sense that he would describe days that began with evening instead of morning.

In a different speculation, it occurs to me that it could be difficult to notice a burning bush on a mountain in the desert while the sun was blazing down. The Bible does not tell us the time of day Moses first noticed the burning bush. Perhaps Moses noticed the bush as the sun slipped over the horizon draping the desert in darkness. In this view, his theophany began in the evening. If so, then his pattern of evening followed by morning might make a bit more sense. Either way, he followed this pattern throughout creation story: *evening followed by morning* the third, the fourth, the fifth and the sixth day. Even now, a new day begins with sundown in Jewish communities.

The days I'm describing were Moses' six days of viewing, not God's six days of creating. Conjecture, to be sure, and we are equally sure the cosmos has reached its present form over billions of years, not in only six, 24-hour days. Through inspiration, whatever form it took, Moses came to know what God said, and then he described what happened in response. Call them Godly quotes and Mosaic commentary, if you please. The quotes admitted no freedom for Moses to choose the words he penned; he just wrote exactly what God said. The rest of what he wrote contained some commentary since Moses chose the words he thought best described those things he saw happen in response to what God said. The exciting thing is, as we compare the evidence-based truth claims of modern science with the inspired truth claims of Genesis, we find them to be synonymous.

After Their Kind

As discussed earlier, Genesis recounts the beginnings of three categories or kinds of life. It is important to notice that God did not create these life forms in a direct, hands-on manner. Instead, God assigned this employment to the earth and to the waters. First, at God's command, the earth brought forth different kinds plants. Second, at God's command, the water brought forth different kinds of animal life. Third, at God's command, the earth brought forth other kinds of animal life.

Some of those who deny evolution point to the *after their kind* to mean that life can only produce what already exists. This view is clearly refuted by science. While it is true that cucumbers do not produce cats, and water melons do not produce water buffaloes, various kinds of life did evolve into various kinds over long periods of time. These clearly produced offspring of the same kind, but over long periods of time, they produced others not identical to the parents. We could say one kind followed another — *after their kind*. The scientific evidence is quite clear on this point; we call it Evolution.

So according to the first chapter of Genesis, God prepared a suitable habitat and then called forth three kinds of primitive life; these are scriptural truth claims. As noted in Part II, Carl Woese taught us there were indeed three aboriginal kinds of life. Obviously, the two truth claims are complimentary. Coincidence? Well, one's Sitz im Leben will have the first say on whether it is coincidence or evidence, but evidence will have the final word. We are confident three kinds of very early life evolved into other kinds of life we see around us today.

Given that the three primitive kinds of life arose through processes which God initiated, we can say the males and females mentioned in Genesis 1:26 were also the products of evolutionary processes. Genesis tells us nothing about how God created these males and females in His image. This is speculation. Perhaps God tweaked the DNA of one of the earlier kinds of hominids in order to make these males and females different from those who came before them.

What is not speculation is that God did something that gave these hominids dominion over the other forms of life. The meaning of the phrase *"in God's image"* remains a subject of debate, but in consideration of the other information in the first chapter of Genesis, it seems that "in God's image" is at least partially, and probably best, explained by the dominion God gave them over the other forms of life. God has absolute dominion over everything, and these males and females were granted some measure of the dominion God possesses without limit. Some may prefer to believe that God spoke and the males and females with dominion suddenly appeared. And to be sure, God could have done these things in just this way. But it seems more plausible to me that God produced these folks through some sort of process.

In this view, the males and females mentioned in the first chapter of Genesis evolved from the more primitive hominids into anatomically modern Homo sapiens, and they went on to become fully modern Homo sapiens due to Divine intervention. Some of these folks did populate the planet as God instructed them to do. As discussed in Part II, we know some of these folks emigrated out of Africa into the Fertile Crescent where they encountered their evolutionary cousins, the Neanderthals. The Neanderthals and the other primitive hominids are now all extinct, clear evidence they were not the ones God gave dominion, for if they had dominion, they would still be around.

Adam did not arrive on the stage of history in the evolutionary manner, and efforts to weave the first two chapters of Genesis into one story about Adam have been, still are and will continue to be exercises in futility. Chapter One says nothing about Adam. He was different. God created Adam in a one-off manner, and God gave him perfect DNA and a soul.

The Traveling Toladah

All great truths begin as blasphemies.

— George Bernard Shaw

The Book of Genesis sets forth the history of the creation of the universe. This biblical outline includes the formation of Earth and of the various forms of plant and animal life. Through science we have determined the cosmos appeared about 13.8 billion years ago (BYA). Earth formed around 4.5 BYA, and life began to show up in the fossil record about 3.5 BYA.

The biblical account is presented in a highly compressed outline, and because it is history, we should not be surprised to find phrases such as: "this is the history of," "these are the generations of," "the genealogy of," "the descendants of," and so on. Behind these phrases is the Hebrew word תּוֹלְדוֹת (pronounced, to-led-**aw**) from the root, יָלַד (pronounced, yaw-**lad**) which means 'to bear' or 'bring forth' or 'beget.' Which of these phrases is used depends upon the version of the Bible one chooses and the context in which it occurs.

For our purposes, we will identify each of these phrases as a *toladah*. Billions of years of begetting and bringing forth have resulted in a long history, and understanding what can be seen in the rearview mirror will help us better understand not only our present circumstances, but also the road ahead.

Toladah first appears in Genesis 2:4-6. This toladah addresses the generations of the heavens and the earth. Other instances of the toladah are found in Genesis 5:1 (the genealogy of Adam); 6:9 (the genealogy of Noah); 10:1 (the genealogy of the sons of Noah); 10:32 (the families of the sons of Noah); 11:10 (the generations of

Shem); 11:27 (the generations of Terah); 25:12 (the genealogy of Ishmael); 25:19 (the generations of Isaac); 36:1 (the genealogy of Esau) and 37:2 (the history of Jacob).

Notice that all these phrases contain the word "of." This small preposition serves to make clear which segment of history the toladah is meant to address before it goes into the particulars. Just below is the first instance of the toladah in the NKJV.

These are the generations of the heavens and of the earth when they were created, in the day that the LORD God made the earth and the heavens, And every plant of the field before it was in the earth, and every herb of the field before it grew: for the LORD God had not caused it to rain upon the earth, and there was not a man to till the ground. But there went up a mist from the earth, and watered the whole face of the ground. (Genesis 2:4-6)

We are very fortunate that the truth claims in this toladah are clearly stated and solidly anchored to specific historical epochs. Call them mile markers, if you please. These mile markers are well known to science.

We know earth was very different in the past than what we see around us today. There were no plants or hominids on early Earth. Actually, the very early Earth didn't host any kind of life. Earth's early atmosphere would have been toxic to most forms of life, including human beings.

It took about one billion years for organisms known as cyanobacteria to convert the noxious atmosphere into one rich in oxygen. Hominids who engaged in agriculture are a recent addition to the historical record.

The particulars of how life emerged are debated, but emerge it did, and life continues in the oxygen rich atmosphere we enjoy today. Thanks to science, we have a pretty good understanding of these histories.

Without doubt, the first verse of this toladah speaks to the creation of the heavens and the earth. It continues by narrowing the focus onto earth in the time before plants. Then, rather oddly, it adds another mile marker indicating this was a time when *there was not a man to till the ground*. Tilling the soil is synonymous with farming, and in the era before plants the idea of farming is nonsensical.

Farming is a relatively recent development in hominid history. As is almost always the case in archaeology, exact dates are unavailable, but evidence of farming shows up in the fossil record about 12,000 years ago in that part of the world known as the Fertile Crescent. The Crescent is a large area of land, encompassing millions of acres, some of which, according to the description of rivers in Genesis 2:10-14, are related to the area known as the Garden of Eden. Furthermore, Genesis 2:15 makes it perfectly clear that God put Adam in the garden to till and keep it. So this toladah (Genesis 2:4-6) is not about Adam. The reader will judge whether or not it is only coincidence that farming shows up in the archaeological record in the region of the world where Adam lived.

The phrase *there was not a man to till the ground* can be understood in two different ways, neither of which involved Adam. The most common interpretation of this phrase is there was no man of any kind, period. Okay, but if there was no man of any kind, then why would the text mention agriculture? Is it not blindingly obvious that if there was no man of any kind, then there would have been no man to till the ground? There would have been no man to hunt or to gather, no man to build fires or to do anything else. Not only so, but the toladah also makes perfectly clear that it addresses the time before plants, and in the era before plants, there could be no agriculture. Once again, why would scripture mention agriculture? So the "no man of any kind" explanation is untenable.

The second interpretation is there was a man present, but not the kind of man who was capable of farming. As we noted in Part

II, the evidence shows a number of other species of hominids occupied the planet before fully modern Homo sapiens came along. There were Neanderthals in Europe, Asia and the Fertile Crescent and anatomically modern Homo sapiens in Africa, but these other hominids were not farmers. They ate some of the plants, but they did not *till the ground*. Instead, they got their calories by hunting and by gathering what grew without their help.

However, this interpretation isn't much help because, as already pointed out, the passage clearly speaks to the time before plants. And once again, if there were no plants to cultivate, the idea of farming would be nonsensical. So the idea this toladah is about Adam and Eve cannot stand scientific scrutiny. At the very least, these comparisons of science with scripture must put the careful reader on notice that something has gone amiss in this passage.

This toladah (Genesis 2:4-6) unequivocally begins with the creation of the universe. It clearly addresses the time before plants and plainly states that during the time of interest there was no man to till the ground. Given the specifics about the segment of history to be covered, the issue of how to understand this toladah is problematic because, in its current location (Genesis 2:4-6), *none* of the history it says it will address actually *is* addressed in the verses that follow it. Instead, the history it says it will address begins in Genesis 1:3, not in Genesis 2:7 as would be expected. So, there is a significant and revealing problem with this passage as traditionally interpreted. This analysis suggests, at least to me, the toladah has been moved. In its present location, what immediately follows this toladah is the story of Adam, but Adam lived after, *not before,* plants show up, and Adam did *till the ground*. In addition, the idea that Genesis 2:4-6 is the toladah of Adam creates a glaring redundancy since the unequivocal toladah of Adam begins in Genesis 5:1.

Adam was different from those who came before him. The earlier hominids did not have the cognitive abilities required to name objects or to till the ground. Adam had both. Two approaches have been employed in attempts to harmonize the discrepancy between what the toladah of 2:4-6 specifically says it will address, and what it actually does. Because this is such a crucial issue, we'll look more closely into what this text actually says, and how we might better understand it. These are not new issues. So let's consider what two prominent theologians had to say about this problematic passage.

The late Reverend John Calvin is well known for his *Institutes of the Christian Religion*. He also wrote commentaries on several books of the Bible, including one on the Book of Genesis. Addressing the issues we've been describing just above, Calvin said, "The term (toladah) signifies, sometimes, the origin of the thing spoken of, sometimes, the posterity of those who are mentioned. It is taken here in the former of those senses; and in chapter 5:1, in the latter."

A different view was given by Keil and Delitzsch in their landmark series, *Commentary on the Old Testament*. According to Keil, the toladah (Genesis 2:4-6) must be understood to only apply to what follows it. So Keil's view was this toladah is about Adam, and this has been the most commonly held view for centuries. Keil went on to disparage those whose opinions differ from his own by saying, "The words, 'these are the (toladah) of the heavens and the earth when they were created,' form the heading to what follows. This would never have been disputed, had not preconceived opinions as to the composition of Genesis obscured the vision of commentators." Keil was correct when he said this toladah (Genesis 2:4-6) must be understood to only apply to what follows it, and he was also correct in his assertion that preconceived opinions can obscure the vision of commentators. Preconceived opinions are part of everyone's Sitz im Leben. However, just because someone disagrees with *our* vision, does not necessarily mean *they* are the ones who suffer from obscure vision. The possibility does exist that it is our vision that has been obscured. Keil argued that a toladah forms the heading for what follows, not summaries of what went before, as Calvin claimed. And Keil's explanation is correct, even though it highlights, but then ignores a glaring problem with this particular toladah.

The problem is even though Genesis 2:4-6 clearly states the segment of history it is meant to address, nothing of this history is mentioned in the verses that follow it. Furthermore, Keil's view that this toladah is about Adam runs aground because Adam lived *after* there were plants on the Earth, not *before*. Also, Adam did till the ground. So, the views of Calvin and of Keil are problematic in different ways, but for the same reason: They worked without the science associated with evolution, a deficiency which led them to the faulty assumption that Adam was the first hominid on this planet which he was not. (Religion without science really is blind as Einstein pointed out.)

Exactly when God created Adam is unknown, but, as noted in Part II, we have unequivocal evidence that other species of hominids roamed about in Africa for the past few million years, and in the Middle East, Europe and Asia during the last few hundred thousand years. These hunter-gatherers left their footprints in the dust long before God created Adam from the dust. Calvin sought to solve the disparity between what this toladah says it will address, and what it actually does, by putting the cart before the horse when he asserted this toladah is meant to address the origin of things spoken of, rather than the posterity of those mentioned. Keil ignored the glaring discontinuity between what this toladah says it will address and what it actually does. Lest we seem overly critical of these two luminaries, we point out again that they labored without the scientific knowledge required to untangled the confusion. Happily, we now have the science required to sort these inconsistencies out, but we'll never have an exact birthday for Adam.

The above facts clearly show this toladah cannot be about Adam, for Adam surely was created after plants came along, and Adam was fully cognitive as is clearly seen in his naming of the animals, which, by the way, had to have plants for nourishment. So, my conjecture is that this conundrum is the result of an editorial change made to support the mistaken idea that Genesis 2:4-6 is about Adam. It is not.

Because our earliest copies of Genesis were written long before the editors knew anything about the Big Bang, paleoanthropology, evolution, genetics, or any other science for that matter, and because more recent commentators were perhaps unable, or unwilling, or maybe even fearful, to directly connect what Genesis tells us with what we have learned from science, the first two chapters of Genesis have withstood previous attempts to solve the problem associated with Genesis 2:4-6. The ruts associated with the idea this passage is about Adam are deep, and as previously noted, new ideas which take us outside the ruts will certainly face J. B. S. Haldane's gauntlet. Those who espouse what is known as "inerrancy of scripture" may assign my conjectures to Haldane's first category, or call them blasphemies, as George Bernard Shaw said would happen. One's Sitz im Leben will have the first say on these matters, but the evidence will have the final word.

Scholars such as Calvin and Keil were not immune to the influence of their Sitz im Leben. Albert Einstein was not immune from the influence of his, and neither is the present author immune from the influence of his. The same observation applies to those reading this book. The evidence suggests Genesis 2:4-6 has been moved from its original location to where it is today. I am not suggesting that the wording of the toladah was changed, only that the editors did the equivalent of what we would call a 'cut and paste.' This well-intentioned, but poorly informed, decision has done more to muddle the message of the first two chapters of Genesis than has anything else we can name. Considering what this toladah says it will address, but doesn't (at least in its present location) we can ask where might the original location have been. Evidence is the best antidote for the Sitz im Leben, and evidence depends on facts. So, what are the facts?

The toladah states, in no uncertain terms, that it speaks to the time when the heavens and earth were created, and the details of the creation begin in Genesis 1:3 where we read that God first created *light*. Therefore, Genesis 1:3 can be taken as the beginning of the timeframe mentioned in this toladah. A second indicator of the 'cut and paste' can be seen in verse 2:5 which expressly states it addresses the era before plants. This indicates that the toladah would have been positioned before Genesis 1:11 for this is the verse that speaks of the beginning of the era of plants. These two considerations support the idea that Genesis 2:4-6 was moved from its original location (immediately following Genesis 1:2) to where it is today.

The idea of early editors moving a sacred text is novel, perhaps even shocking to modern Christians, at least to those unfamiliar with the documentary hypothesis proposed by Wellhausen. In order to consider this idea would require confronting our own Sitz im Leben. We can do so by jumping, at least temporarily, outside the ruts. In order to test this conjecture, we turn to the evidence.

First, let's consider how this decision to move Genesis 2:4 may have come about. The earliest manuscripts of Genesis did not include chapter and verse numbers; these were added by later editors. We appreciate and very much need these numbers to help us quickly find the information we seek, for without the numbering it would be a daunting task to find a given passage in a timely manner. However, as useful as they are, the numbering of chapters and verses creates a somewhat artificial structure, one that would argue against moving a passage from one spot to another, even if the words of the passage were not changed in the slightest. There does exist a firm chronology in much of the biblical narrative because most of the text is history, and obviously, history takes place with the passage of time. So some passages of scripture would not permit moving the text. For example, moving the account of the birth of a child in such a way as to make the child's birth come before the birth of the child's mother would be ridiculous.

However, not every passage of scripture is so strongly tied to an immovable timeline. It could have been that the early editors who worked with the unnumbered verses were willing to move a passage around to make the text more comprehensible, at least in their views, which as already pointed out, didn't incorporate a lot of science. This is speculation. We cannot be sure about these editorial moves, even given the scholarship of the biblical critics. We cannot be sure because we do not have the Mosaic autograph. However, if Julias Wellhausen and his colleagues were correct in their documentary hypothesis, we can be sure that some passages were moved about as sources were redacted, combined and edited into drafts that were eventually compiled into a final version.

To summarize my conjectures, the first toladah of the Bible (Genesis 2:4-6) has been moved; it was originally located immediately following Genesis 1:2. I am not suggesting the editors changed the wording of the toladah. This 'cut and paste' move was probably made because the editors knew nothing of the hominids who lived long ago, a deficiency that led them to believe Adam was the first man on this planet, and therefore, this toladah had to be about Adam. The problem was that in its original position (Genesis 1:3-5), it clearly was not about Adam, for Adam did till the soil. However, they knew, as Keil pointed out, a toladah is meant to only apply to what follows it. It seems, at least to me, that they moved it in order to force it to apply to Adam. Whether it fits well with one's Sitz im Leben or not, Genesis 2:4-6 clearly speaks to the time *before* plants and *before* Adam, *not* the time of plants and of Adam. Not only so, but as mentioned above, the toladah of Adam (his genealogy) unequivocally begins in Genesis 5:1.

Another reason the first six verses of Chapter Two of Genesis are jumbled is the editors who added the chapter and verse numbering divided the texts into chapter one and chapter two at the wrong place. Genesis 2:1-3 is clearly a summary of events described in the first chapter of Genesis, and summaries don't open chapters; they close them.

Here is the sequence of passages that constitute chapter one described in my conjecture.

- Genesis 1:1-2............The Headline
- Genesis 2:4-6............The toladah
- Genesis 1:3-31.........Details promised in the toladah
- Genesis 2:1-3...........Summary of chapter one

In this view, Genesis 2:7 would be the first verse of the second chapter of Genesis.

DISCLAIMER

The conjectures just above do not alter the truth claims connected with the decision which will determine where the soul goes when a person dies. The truth claims remain unchanged. Only their relative positions within the texts are questioned. The conjectures are meant to untangle the first two chapters of Genesis and thereby clarify the connections between science and scripture they contain.

Concerning the Sabbath

Thus the heavens and the earth were finished, and all the host of them. And on the seventh day God ended his work which he had made; and he rested on the seventh day from all his work which he had made. And God blessed the seventh day, and sanctified it: because that in it he had rested from all his work which God created and made. (Genesis 2:1-3)

If the time required to complete a given task is proportional to the difficulty in doing it, then bringing the cosmos into being was the easy part. It took only enough time for God to say, "Let there be light." The task of shaping the cosmos into what we see today has taken much longer. The tiny cosmos was created in an instant, but it has taken almost 14 billion years for it to reach its present configuration.

It took roughly the first 9.3 billion years for the Earth to evolve from its *without form and void* condition into a rocky, oblate spheroid orbiting a star. It took roughly a billion years more for the first three kinds of very primitive life to emerge and about another 3.5 billion years for anatomically modern Homo sapiens to come along. Given the time required for all this to come about, we can say that it was no small feat, even for God. Little wonder God took a day off. Right?

Well, the scripture above is not telling us that God was fatigued and did nothing more. Instead, Genesis 2:1-3 summarizes chapter one and lays the foundation for Sabbath. Human beings do get tired; we do need time to rest, to worship and enjoy family relationships.

There was a time in America when the so-called "blue laws" made it illegal for non-essential businesses to do commerce on the Sabbath. Some businesses still close their doors on Sunday, but for the most part, those days are gone. Some might argue the country is worse off as a consequence, but I digress.

Life On Other Planets

Advances in science have shown that some interpretations of scripture are no longer tenable. And quite frankly, the faithful, whether they care to admit it or not, are indebted to science for bringing these issues to light. No one wants to hold as true anything that is false. Some faulty interpretations of scripture were embraced because they seemed plausible at the time, and there was no reason to believe an alternative interpretation might be warranted, let alone available.

For example, for most of our history we did not know there were other stars with planets in orbit. This lack of information led to the idea the Earth was the center of everything and that life exists only on this planet. Challenging this dogma caused Bruno, Copernicus, Galileo and others a great deal of trouble. We have learned there are trillions of other stars throughout the universe, and many of these stars have planets. We have not yet found life on other worlds, but it seems, at least to me, we will. Is it possible that life exists only on one planet orbiting a star in a galaxy that contains billions of other stars, a galaxy that is one among billions of others? Well, yes. It is possible, but very unlikely.

As a result of our increased scientific understanding of the Cosmos, the idea that the Earth is the center of everything has been abandoned. We do not know for certain whether there is life on other worlds, but scripture makes clear truth claims about life on this one, and science has been increasingly useful in verifying these claims.

As noted earlier, the first chapter of Genesis states that certain kinds of animal life started in the waters, and although the details are debated, this biblical truth claim is affirmed by science. In Genesis 1:20 we read, *And God said, Let the waters bring forth abundantly the moving creature that hath life, and fowl that may fly above the earth in the open firmament of heaven.* No doubt, this passage refers to life on Earth, but is this the end of the matter?

If there are other worlds that have water, this begs the question of whether or not God's command to the waters to bring forth life pertained only to the waters on Earth. We have no evidence of life on other worlds, at least not yet, but my Sitz im Leben includes the idea there is life on many other worlds where water exists. For the present, it is enough to recognize that science and scripture claim that some early kinds of animal life began in the waters. We can't say precisely where, or exactly when, but for people of faith, the first kinds of animal life began as the result of God's command that the waters *bring forth*. Christians believe the life that came forth from the water did so in response to God's instruction. Even though neither science nor scripture provide the particulars of how the first life came to be, the two agree that water was a primary assembly facility, if you will.

Unless we believe that only the water on Earth obeyed the command of God, we should expect to discover life on other worlds that have water. Lest we join the ranks of those who bash our species for a variety of imperfections, notice in this case, our failure to recognize the possibility of life on other worlds was not the result of closed minds or personal agendas. In this case, we had not decided, a priori, what the answer to the question of life on other worlds would be. Our ancestors didn't even realize there was a question to be posed on this issue. Today, the question of life on other worlds has become mainstream, and we have built expensive and sophisticated instruments to try and settle the question of whether or not we are alone.

So, God told the waters (not ice or steam) to bring forth animal life. As high school chemistry students know, the symbol for hydrogen is H and the symbol for oxygen is O. So, when two atoms of hydrogen combine with one atom of oxygen they form one molecule of water, which we write as H_2O. As noted in Part I, hydrogen was formed during the early stages of the Big Bang, but oxygen was produced inside stars millions of years later. The two gases combine to form a precious liquid, the elixir which is absolutely necessary for life as we know it. As is also well known, H_2O freezes at zero degrees Celsius and becomes steam at 100 degrees. Obviously, liquid water requires a temperature between zero and 100 degrees Celsius (32 and 212 degrees Fahrenheit).

We also understand that temperature depends on the distance between the planet and the star it orbits, as well as on the star's energy output. In order to have liquid water, planets which orbit hotter stars must be further away from their star, and those orbiting cooler stars must be a bit closer to theirs. Scientists sometimes call this special orbit the Goldilocks Zone. Such an orbit allows the planet to be not too hot and not too cold, but just right. Obviously, the Earth is located in one of these goldilocks zones.

For generations, it was assumed the Earth was the only planet, period. Prior to the twentieth century, few would have imagined the cosmos contains some hundred billion galaxies, which it surely does, and galaxies contain some hundred billion stars, which they clearly do. Given the number of other galaxies, it is a virtual certainty that large numbers of other planets with water are out there, and they very well may have living organisms aboard. Why? Because God said, *"Let the waters bring forth,"* and there is no indication that only the waters on Earth responded to God's command. But wasn't God speaking only to the waters on Earth? That's been the most common assumption for centuries, and it could turn out to be the case. However, the cosmos is a staggeringly huge place, and it may be that God's plan for life is more extensive than we previously assumed. Maybe, just maybe, life does exist on many other planets where conditions allow for liquid water.

What is certain is the water on Earth did bring forth *living creatures* and *birds*. It seems unlikely that birds popped out of the water like bread out of a toaster. The fossil evidence supports the view that the *bringing forth* took a long time and included intermediate kinds of life. We have named these processes Evolution, but let's be clear. Christians believe God caused the water to bring forth life. So even with the dawn of the scientific age, the most common assumption was God's command to the water to bring forth life was limited to only the water on the Earth. Obviously, the Bible is an earth-centric book, but to assume the evolution of life from the water applies only to Earth, may be a step too far. Given the vast expanse of the cosmos and God's specific command to the water, we should not be surprised to find various kinds of life in many other watery places — out there.

Thus far, in Part III we have compared the Genesis account of creation with what we have learned from science, and from these comparisons we concluded that Genesis contains a good description of the history of the universe from the perspective of one viewing from the outside. We argued that through inspiration, God provided the information necessary for Moses to write, and we speculated how that inspiration might have taken place. We argued that early editors moved Genesis 2:4-6 from its original position in order to force it to apply to Adam, which it was not originally meant to do. We pointed out along the way that one's Sitz im Leben would certainly come into play as we try to get out of the theological ruts deepened over the centuries. Without the comparisons of scripture with science, getting out of the ruts would be very difficult, if not downright impossible

Rightly divided, the second chapter of Genesis begins in verse seven with the creation of Adam:

And the LORD God formed man of the dust of the ground, and breathed into his nostrils the breath of life; and man became a living soul. (Genesis 2:7 — KJV)

We'll take up the particulars of this very important story about Adam after we discuss the setting.

The Garden In Eden

Mythologies can begin with phrases such as, "Long ago in a land faraway" or "On a strange planet in a distant galaxy" or something else along these lines. Genesis specifies the locale of the events of interest, and it occurs to me that Genesis provides such detailed information, not only because these are the lands where these events actually took place, but also to assure the reader that Genesis is history, not myth. Still, we cannot specify the exact location of the Garden of Eden or the exact time Adam and Eve lived there. However, a number of key events which led to the arrival of fully modern Homo sapiens took place sometime after about 50,000 years ago in the part of the world called the Fertile Crescent.

The Fertile Crescent is the name assigned to a loosely described, large swathe of land running from modern Egypt, through Israel, Jordon, southern Turkey, Syria, Iraq and Iran. This area is also called the Cradle of Civilization, for it is in this part of the world that activities we associate with fully modern Homo sapiens begin to show up in the archaeological record. If we zoom in from high above the northern portion of the Fertile Crescent during the time of Adam, we would see a smaller parcel known as the land of Eden. If we zoom in once more on the eastern portion of Eden, we would see a smaller area covered with splendid trees. God planted this arboretum in Eden and put Adam in the garden to till and keep it. It is clear from Adam naming the animals and his tilling the land that he was fully modern in his cognitive skills. Here is the text that describes Adam's first home. (Once more, all biblical texts are from the NKJV unless noted otherwise.)

And the LORD God planted a garden eastward in Eden; and there he put the man whom he had formed. And out of the ground made the LORD God to grow every tree that is pleasant to the sight, and good for food; the tree of life also in the midst of the garden, and the tree of knowledge of good and evil. (Genesis 2:8-9)

So, the Lord God planted this garden in the eastern portion of Eden with trees that were *pleasant to the sight* and *good for food*. Very importantly, there were two trees in the garden that were different from the others: *the tree of life* and *the tree of the knowledge of good and evil*. A hint that there is more to this story than first meets the eye can be seen in the fact that only these two trees had names. Furthermore, and contrary to the usual assumption, there is no textual support for the idea that God planted either of them. Actually, the text only says that these two were there among the others. My conjecture is that the *tree of life* and *the tree of knowledge of good and evil* were not ordinary trees. Neither had roots in the ground or branches in the air. We'll come back to these two extraordinary trees a bit further down. Skeptics consider the story of the Garden of Eden to be nothing more than mythological folklore, but the detailed description of the land of Eden suggests otherwise. Here is the relevant text.

Now a river went out of Eden to water the garden, and from there it parted and became four riverheads. The name of the first is Pishon; it is the one which skirts the whole land of Havilah, where there is gold. And the gold of that land is good. Bdellium and the onyx stone are there. The name of the second river is Gihon; it is the one which goes around the whole land of Cush. The name of the third river is Hiddekel; it is the one which goes toward the east of Assyria. The fourth river is the Euphrates. (Genesis 2:10-14)

Many agree that two of the rivers in this passage are the Tigris and Euphrates, but the identity of the other two is uncertain. The

Tigris and Euphrates are big rivers, so the land of Eden was no small place. As explained in Part II, we have unequivocal evidence that different species of hominids lived in this part of the world during the last 100,000 years, and even though these roaming bands of hunter-gatherers were few in number, their actions profoundly shaped hominid history. The Garden of Eden was home to some of our ancestors, and a growing body of archaeological and genetic evidence has shown our history is intermingled with the history of others who walked among the trees and drank from the rivers of Eden.

Diversity In The Garden

In order to appreciate the events that took place inside the Garden of Eden, we need to understand, as best we can, the characters who lived in and around the area. Unequivocal evidence shows that Neanderthals and anatomically modern Homo sapiens, as well as hybrids of these species, lived in and around the Fertile Crescent in the relatively recent past. Israel Hershkovitz pointed out that Neanderthals and humans were living in Manot Cave 55,000 years ago.[49] Adam and Eve would come to reside in the garden, but they were not the only residents. We read in Genesis 1:26-29 that God had previously created other males and females, blessed them and sent them out into all the world. Some of these blessed folks were in the land of Eden.

The exact nature of the blessing is unknown, but whatever it was, it changed the recipients in some way. We must assume the change was an improvement of some sort. Obviously, the blessing was bestowed on creatures who were there to receive it. So what might have been the mechanism that brought the blessing into effect?

Since DNA orchestrates our physiology and cognitive abilities, I'm entertained by the idea that God blessed them by tweaking a portion of their DNA. This is speculation. Perhaps God was not pleased with the way things were going and decided to intervene to move things along a bit differently. In this view, rather than call forth an entirely new line of evolutionary hominids, God did the equivalent of what we call a CRISPR/cas9 gene edit to enhance the DNA of those already present. If so, God altered their genetic code in some manner that gave these males and females enhanced abilities the other animals did not possess, abilities that would result in them having dominion over the others.

This does not mean the entire population of anatomically modern Homo sapiens woke up the next morning to discover they had the complete toolkit necessary for fully modern behavior. Although the new potential was put in place quickly, its full implementation involved a process.

Those who were blessed when God tweaked their DNA quickly acquired the body parts necessary for fully modern behavior, but in the story I'm telling, the new potential, which Professor Ian Tattersall calls the "keystone," lay fallow until the new owners discovered how to use it. Rather like dry kindling needs a spark to get the fire started, the new potential needed some sort of cultural stimulus to ignite the cognitive fire. So what was the cultural stimulus? Well, God created Adam and put this man with perfect DNA in the mix. My conjecture is the cultural stimulus was their interaction with Adam.

As their numbers grew and travels increased, some of them almost certainly passed through the Garden of Eden. These hominids were anatomically modern Homo sapiens, and even though they had lived for centuries as creatures of instinct, they were not stupid — especially after the Godly blessing.

Africa was their homeland, and many of the other animals on the dark continent were stronger and faster and equipped with large teeth and sharp claws. Various kinds of African carnivores wanted to have these early hominids over for dinner, not as guests around the table, but as the main course on the table. Weaker individuals (the elderly, the sick and the injured or those who could not run as fast as other members of their tribe) were more often the victims of predators than were the stronger, faster, more agile folks. If survival of the fittest really is a factor in evolutionary history, those who survived attacks by predators tended to be faster on their feet than those who became prey, and as a consequence, faster members of the tribe tended to live longer and have more children than the slower members of the tribe.

In their struggles for survival, these early hominids learned to be cautious and stealthy when the need arose. I imagine when they first saw Adam in the Garden of Eden, they watched him from concealed positions, at least until they were convinced that Adam was not a threat. With the passage of time, they learned from Adam that food could be acquired by growing plants in the soil. This is valuable information because hunting wild animals with primitive weapons is very hazardous business. They also learned that objects can have names, and as stated above, this realization is the keystone to complex language. In this view, their interactions with Adam were the stimuli that thrust them into fully modern behavior.

Since the hunter-gatherers lived in small tribes, the cultural stimulus started out by affecting only a few individuals, but the stimulus was so powerful that those affected would never again see the world in the same ways. There were not large numbers of people during these times, no more than a few ten thousand, and they lived in small groups, scattered over large areas. So new ideas spread slowly across the larger population. This means the cultural stimulus was not a one-off event that impacted the entire population within a few hours. Chance encounters between the scattered groups of these nomadic hunters may have been infrequent, but they did occur. When different tribes crossed paths, the effect of the blessing, coupled with the cultural stimulus that activated it, spread from one to the other. It would have taken time for the new way of processing information to become wide-spread.

Even so, due to the blessing of God, when they were exposed to the cultural stimulus, they were biologically prepared to receive it and able to employ it. Through their interactions with Adam, the anatomically modern Homo sapiens transitioned from non-symbolic, non-linguistic, hunter-gatherers, to fully modern Homo sapiens who named objects, communicated in complex ways and engaged in agriculture. My conjecture is these interactions with Adam were the cultural stimuli that Professor Ian Tattersall does not identify, but very convincingly argues, must have taken place in order for anatomically modern Homo sapiens to make the changeover to fully modern behavior.

As their numbers grew and the cognitive changeover became wide-spread, the other hominids, those who did not receive the Godly blessing, were no longer able to effectively compete. All the other species of hominids, including Neanderthals, are extinct.

God is sovereign. God blesses anyone or anything He chooses, and He does so wherever, whenever and however He wants. Since the code of life is written in DNA, the species with superior DNA will inevitably come to dominate the others, for in the long struggle for survival, keen wits are more effective than brute strength. The other animals, including the other hominids, came onto the stage of history through the long process we call Evolution, and I have conjectured that God tweaked the DNA of one of the primitive species to give them dominion over other forms of life. Even so, their new potential lay dormant until they began to interact with Adam. We can only speculate as to why God did things in this manner, but, as mentioned earlier, God is sovereign.

Adam and Eve

The most interesting stuff in hominid history took place in and around the Garden of Eden, for this is where God put Adam and

Eve, and this is where fully modern Homo sapiens made their way onto the stage of history. Adam and Eve and the serpent had speaking parts, but there were other characters in the Garden who did not speak, or if they did, we have no record of what they said.

However, and this is crucial to understanding the story, even though God wrote the script, He allowed Adam and Eve a great deal of freedom in how they played their parts. This very important privilege is what the church calls "free will." So for the present, we are most interested in those characters who played their parts in and around the Garden of Eden, and due to advances in science, we are now in position to identify and better understand the actors in this amazing story and how their actions shaped the history of modern people.

The Man With No Navel

Adam was different from the other hominids; he was not born of woman. God formed this man in a one-off fashion. Once again, here is the beginning of Adam's story.

And the LORD God formed man of the dust of the ground, and breathed into his nostrils the breath of life; and man became a living soul. And the LORD God planted a garden eastward in Eden; and there he put the man whom he had formed. (Genesis 2:7-8, KJV)

As mentioned earlier, we must assume that Adam had perfect DNA, for the idea that God put faulty genetic code in Adam is simply beyond the pale. The text does not mention Adam's initial configuration. We do not know whether Adam was tall or short; whether he was hairy or smooth skinned; whether he had brown eyes or blue. We cannot say whether Adam was a teen, a young adult, or a mature man when God put him in the Garden of Eden. But the lack of this sort of information will not much hinder our work.

It would not be incorrect to say that Adam was the first fully modern Homo sapiens, but it would be incomplete because, at least in some ways, Adam was superior to people living today. For example, modern people have to teach their children what a dog or a ball is, but Adam knew that objects (animals in this case) can have names, and he knew how to speak those names. We have no way of knowing what language Adam spoke.

Likewise, there is no record of anyone teaching Adam how to till the soil. Not only did Adam's DNA give him superior cognitive skills, it enabled him to live far longer than the oldest members of modern society. According to the toladah of Adam, he was 130 years old when he fathered a son named Seth, and Adam lived a total of 930 years![50] God created Adam with the cognitive abilities seen in modern people, and Adam had some extraordinary abilities that are long gone. Adam named objects, tilled the soil in the Garden and lived a long time because God gave him perfect DNA.

Even though Adam had no earthly parents or other family, he was not alone, not even before God cloned Eve from Adam's rib. The other hominids in the area became aware of Adam, and as their confidence grew, they began to interact with this man without a navel. Through Adam, the anatomically modern Homo sapiens learned how to use the enhanced abilities that God gave them through the blessing. These newly actuated abilities may have been the deciding factor in the extinction of Neanderthals. A number of these folks who meandered through the Garden of Eden had made changeover to fully modern behavior before God cloned Eve from Adam's side. Although Adam was not alone in the sense that other hominids passed through the Garden of Eden from time to time, he was alone in the sense that none of the other hominids were his equal.

Adam had almost certainly seen the various kinds of animals having sex. Not only so, but in order for the other hominids to fill the Earth as God instructed them to do, they had to have lots of children, and Adam very likely had also seen some of these folks engaging in sexual behavior. We can say the information exchange was a two way street. The cognitive awakening in the anatomically modern Homo sapiens was triggered by their interactions with Adam, and Adam's growing interest in sex was fueled by his observations and interactions with his nomadic visitors.

So the Garden in Eden was the setting for the events of interest. And the man with no navel was not alone, not even before God brought Eve onto the stage of history.

The Birds and The Bees

Following the description of Eden and the creation of Adam, the narrative turns to a certain restriction God imposed on him.

And the Lord God commanded the man, saying, "Of every tree of the garden you may freely eat; but of the tree of the knowledge of good and evil you shall not eat, for in the day that you eat of it you shall surely die." And the Lord God said, "It is not good that man should be alone; I will make him a helper comparable to him." (Genesis 2:16-18)

There is more to this passage than casual reading suggests, and in order to appreciate its genius, we turn to an expert in the Hebrew language, for in any field of investigation, some individuals manage to see further than their peers. In order to better understand the events that took place in the Garden in Eden, we turn to a Jewish scholar who was closer to the ancient text than we are today, and presumably more familiar with the meaning of some old

Hebrew words and literary techniques than are modern commentators.[51]

Rabbi Moses ben Maimon, known as Maimonides (1135–1204) was a physician, philosopher, Old Testament scholar and prolific writer. Maimonides stands out as one of the most influential figures in the history of Hebrew scripture, especially the first five books of the Old Testament known as the Pentateuch.[52] It seems that Maimonides was concerned for members of the Jewish community who were troubled by sectarian philosophy that was on the rise at the time. Apparently, some of the faithful were having doubts about the Jewish Faith as secular reasoning seemed to conflict with Jewish scripture. In the following, we are primarily interested in his volume entitled, *Guide for the Perplexed* in which Maimonides explains that Hebrew is a Holy language. He said, "The Hebrew language has no special name for the reproductive organs in females or in males, nor for the act of procreation, nor for semen, nor for excreta. The Hebrew lexicon has no original terms for these things, and describes them only through figurative allusion and hints."[53] For example, when the scripture uses the phrase, *knew his wife*, it means the man and woman had sexual intercourse.[54] So according to Maimonides, when the story to be told involved indelicate topics, Hebrew authors turned to figurative allusion and hints to carry the content.

Earlier, we suggested that *the tree of knowledge of good and evil* was not an tree with roots in the ground. With Maimonides' insights in mind, a careful reading of Genesis suggests that eating from *the tree of the knowledge of good and evil* is not a story about a what constitutes a healthy diet. Instead, it's an allegory about unhealthy sex. One does not have to be as brilliant as Maimonides to understand that although hummingbirds and honey bees are real creatures, stories about the birds and the bees are allegories about sex.

Circumstantial evidence that this story is about inappropriate sex can be seen in the sequence of the events described. We read in verses 2:16-17 that God told Adam not to eat from *the tree of knowledge*, under penalty of certain and immediate death, and in the very next verse, God said, *"It is not good that man should be alone; I will make him a helper comparable to him."* The prohibition against eating from *the tree of the knowledge*, followed immediately by God's declaration that it was *not good for Adam to be alone*, is not a coincidence. Instead, this sequence of events describes Adam's problem followed by God's solution for the problem. The problem was that Adam was considering getting involved in what he saw others doing, and God's solution was the creation of Eve.

Testosterone is a powerful hormone, and the texts seem to suggest that Adam was feeling the growing influence of his. Sex will find expression in some manner or the other, but regardless of hormonal pressures, not all kinds of sex are appropriate. We can say the time had come for Adam to get married and settle down, but Adam was not meant to mate with the anatomically modern Homo sapiens, not even with those who had made the changeover to fully modern behavior. It was not that God intended for all kinds of sex to be off limits to Adam. After all, there was a reason that God created Adam as male and Eve as female. Obviously, sex is the mechanism for the propagation of the species, and human beings can take no credit for this idea. We did not invent sex; God did, and God knows which sorts of sex are appropriate and which are not. Rather than forbidding Adam to engage in sex of any kind, God's warning was meant to protect Adam from the consequences of inappropriate sex with the evolutionary line of hominids, and the warning is immediately followed by God's declaration that it was time for Adam to have a suitable mate. To be clear, my conjecture is *the tree of the knowledge of good and evil* is an allegory about inappropriate sex, and God's solution for Adam's dilemma was the creation of Eve. God's warning was not a capricious threat; neither was it a test to see whether or not Adam would obey. Instead, it was a warning of what would happen if Adam yielded to the influence of his testosterone.

God's warning to Adam can be compared to that of a loving parent warning a small child to stay away from the chemicals in the cabinet under the kitchen sink. To the child, these brightly colored liquids might be a delight to the eyes, but the parent would want the child to stay away from them. Adam could not have understood what would happen if he ate the forbidden fruit, any more than a child could understand what would happen if he or she came in contact with the pretty liquids under the sink. God knows how things work, and just as a parent knows their child must not drink the pretty liquids under the kitchen sink, God knew what the forbidden fruit would do to Adam and his descendants.

In this view, the time had come for the man with perfect DNA to have a suitable mate, a female who was not only biologically compatible, but also his cognitive equal. God remedied Adam's biological isolation when He cloned Eve from one of Adam's ribs.

As pointed out in Part II, gender is determined by two chromosomes, one inherited from the father and the other from the mother. Males have one X chromosome and one Y chromosome. Females have two X chromosomes. If a child receives a Y chromosome from the father, then the child will be male. Otherwise, the child will be female. It's the father's contribution that determines a child's gender. God gave Eve two copies of the X chromosome. If this was the only change God made in Adam's genome when He cloned Eve, then Adam and Eve were male and female twins.

Upon seeing Eve for the first time, Adam said, *"This is now bone of my bones and flesh of my flesh; She shall be called woman because she was taken out of man."* What an extraordinary meeting this must have been! His response lends credence to the idea that Adam was physically different from the other hominids. When Adam looked at Eve, he saw a female who was different from the other females he had seen. He knew Eve was meant to be his mate.

As far as we know, there were no mirrors in the Garden of Eden, which means Adam did not know what his own face looked like. Yes, he knew what his arms, legs, feet and front part of his torso looked like, but as is the case today, without a mirror, people do not know what their own face looks like. There were obvious and substantial differences between Adam's physical appearance and the appearance of the anatomically modern Homo sapiens around him. If not, how was it that Adam could have immediately understood that Eve was very different from the other females in the area?

Perhaps Adam was significantly taller than the others. Maybe he had light skin, blonde hair and blue eyes. Physical differences between Adam and his neighbors are unknowns, but his comments upon seeing Eve show he immediately understood that Eve was not just another female among the others. She was *bone of his bone and flesh of his flesh*. Adam already had some interesting neighbors, and with the arrival of Eve, he had a suitable mate.

To tidy up just a bit, my conjecture is God blessed a number of anatomically modern Homo sapiens by tweaking their DNA in a manner that gave them the potential to exercise dominion over other forms of life. However, they needed some sort of cultural stimulus to awaken their enhanced cognitive potential. The stimulus came through their interactions with Adam. Even after the change, these folks still carried the genetic mutations their ancestors had accumulated over the millennia, but Adam's DNA was perfect and powerful. Adam's DNA not only gave him superior cognitive abilities, but also enabled him to live much longer than people today. Adam lived over nine centuries, but we do not know Adam's stage of biological development when God put him in the Garden. We do not know the color of Adam's eyes, the color of his skin or how tall he was. By today's standards, he could have been very tall; we just don't know. Neither do we know how long Adam remained a bachelor before God cloned Eve from

Adam's side. Notwithstanding, we do know Adam was in the Garden long enough to name the animals before Eve arrived.

Without question, the toladah of Adam begins in the fifth chapter of Genesis where we read that Adam was 130 years old when he became the father of Seth. My best guess is that the birth of Seth was not too distant from the birth of Cain and Able. In this view, Adam was over 100 years old when God evicted him from the Garden of Eden. Regardless of Adam's stage of biological development when God put him in the Garden of Eden, at the very least, Adam was a fully modern Homo sapiens as is clear from his naming of the animals and his tilling the soil. With the passage of time, Adam became acquainted with some of the evolutionary hominids, and they with him. From Adam, the anatomically modern Homo sapiens learned that objects could have names. This heightened cognitive ability gave these folks the ability to exercise dominion over other kinds of life. The creation of Eve set the stage for what was intended to become a loving, monogamous, husband and wife relationship, and like all intimate relationships, people need time to get acquainted.

Men and women don't start having children immediately upon their first meeting, and even though Eve was created specifically to be Adam's wife, the Garden couple needed a bit of time to establish their relationship before they started their family. In short, Adam and Eve were courting.

Free Will

Adam and Eve were biologically close enough to the others to have children with them, but God did not intend for them to do so. However, God chose to not control each and every aspect of their behavior. The choice of whether to follow God's instructions or their own inclinations remained with the individuals. Adam and Eve were not biological robots; God gave them free will, a power which improperly used, opens the door to major problems. It did in the Garden of Eden, and it has the same potential today. Adam and Eve were created for each other, not for the other hominids in the area.

The anatomically modern Homo sapiens in the area were swept up in a mind boggling cognitive change. They entered the Garden of Eden as creatures of instinct who lived pretty much as their ancestors had done for thousands of years. God had already enhanced their DNA, and through their interactions with Adam, they learned how to exercise dominion over other creatures. In short, they transitioned into fully modern Homo sapiens.

They became what we might call "street wise." Still, they found themselves living in a sort of twilight zone. They would never be exactly like Adam who had perfect DNA, but due to their heightened cognitive abilities, they were no longer like those who came before them. One of these guys is described as being more cunning than any beast of the field which the Lord God had made. He was not naive when it came to sex, and therein was the problem.

Reproductive Choices

One of the anatomically modern Home sapiens males who had made the changeover to fully modern behavior wanted to have sex with the new girl in the Garden. This amorous fellow is portrayed as the *serpent* in the allegory of *the tree of the knowledge of good and evil*. Here is the text.

Now the serpent was more cunning than any beast of the field which the Lord God had made. And he said to the woman, "Has God indeed said, 'You shall not eat of every tree of the garden'?" And the woman said to the serpent, "We may eat the fruit of the trees of the garden; but of the fruit of the tree which is in the midst of the garden, God has said, 'You shall not eat it, nor shall you touch it, lest you die.'" Then the serpent said to the woman, "You will not surely die. For God knows that in the day you eat of it your eyes will be opened, and you will be like God, knowing good and evil." (Genesis 3:1-5)

The *beasts of the field which the Lord God made,* were those who had traveled the long road of evolution, including those who walked about on four legs and those who walked on two. These were creatures of instinct, not symbolic thinkers or moral agents. Even though morality was not part of their ethos, sex certainly was.

My conjecture is the *serpent* was one of the atomically modern Homo sapiens who had made the changeover to fully modern behavior through his interactions with Adam. Not only so, but he could have been under the influence of Lucifer, the father of lies. After his cognitive awakening, the serpent was *more cunning* than the others, and when Eve came along, the serpent used his heightened cognitive skills to strike a conversation with Eve by asking her a question that mischaracterized what God had said. His question was a pickup line meant to get the conversation started. In saloon slang, the serpent was hitting on Eve.

As is well understood, consensual sex requires agreement of the minds of the participants before the physical act takes place. Otherwise, the act is called rape. This story is not about rape. Instead, it's about consensual sex that God had warned against. When Eve came along, the serpent used his heightened cognitive abilities to tempt Eve to do that which God had forbidden. His intention was to seduce Eve, and we can be sure that when the temptation took place, the serpent was *not* a snake. Snakes don't talk, and those who do not talk, cannot tell lies. The serpent did speak, and he did lie. His anatomy was very different from that of a king cobra or a black mamba, as we will show a bit further down. According to the passage just above, this cunning, handsome fellow used a well-crafted lie to get into Eve's head.

Eve was at a sore disadvantage, not only because she was a naked, naive virgin, but also because she had no mother, older sisters or girlfriends to warn her about talking to handsome strangers. What's more, she was operating on the basis of hearsay, for God told Adam not to eat from the forbidden tree before He cloned Eve from Adam's rib. So Eve did not hear for herself what God said to Adam. As has been pointed out many times in other places, Eve added to God's prohibition with her words, *"nor shall ye touch it…"* These words were not part of what God told Adam, and when Eve added them, she inadvertently laid the trap the serpent would use to ensnare her. Just below we read Eve's responses to the serpents advances.

And when the woman saw that the tree was good for food, and that it was pleasant to the eyes, and a tree to be desired to make one wise, she took of the fruit thereof, and did eat, and gave also unto her husband with her; and he did eat. And the eyes of them both were opened, and they knew that they were naked; and they sewed fig leaves together, and made themselves aprons. (Genesis 3:6-7 — KJV)

Eve made three observations that led her to eat the forbidden fruit. First, *the tree was good for food;* second, *it was delight to the eyes;* and third, *the tree was desired to make one wise.* Her observation that the tree was good for food does not mean Eve simply took a closer look and somehow realized the tree was good for food. Instead, something happened that changed Eve's view, and whatever it was, it led her to the wrong conclusion. My conjecture is that during his rebuttal to Eve's arguments against doing what he wanted, the two of them made physical contact. If they touched, even without her permission, and since she didn't die, as she herself had said would happen, then Adam's warning to her was incorrect. The tree really was good for food. Eve and the serpent made physical contact, and when they touched, it started a chain of events that changed history, changed it for the worse and changed it permanently.

Here we see the dangers inherent in adding to what God really said. A powerful lesson to be learned from these events is this: We must find out for ourselves what God has said, and then be careful not to add our own words to those spoken by the Lord. Our beloved spouses, respected colleagues, close relatives and even clerics, all acting with honorable intentions, might lead us to believe God said something which He did not say. So the touch was sufficient proof to Eve *that the tree was good for food.*

Eve also thought the tree was *a delight to the eyes.* In Eve's estimation, the serpent was a good looking guy. Not only was he a cunning fellow, he was, as the saying goes, a handsome devil.

In addition, the serpent used his cunning to appeal to Eve's vanity by telling her if she would do what he wanted, her eyes would be opened and she would *be like God, knowing the difference between good and evil.* Notice that she did *not* know the difference at this point. Neither did Adam.

Tripped up by her own words, and persuaded by the seducer's good looks and lies, Eve formed her opinions and decided to disregard the warning. Despite what Adam had told her, she thought her observations were accurate, her opinions correct and her actions appropriate. *She took of its fruit and ate.* Eve had sex with the serpent, and as the serpent said would happen, she certainly grew wiser, but the wisdom she gained was not a source of pride as the serpent had told her it would be. She came to know the difference between good and evil, and she was ashamed of what she had done.

Thereafter, Eve was no longer a naive virgin, but she was still a woman with no mother, no sisters and no girlfriends to help her manage the mess she had gotten herself into. The mess was not only spiritual; it was also physical.

Disease

We have learned through science that there are dozens of different kinds of human papilloma viruses (HPVs) which can be transmitted through sexual contact. Many of these viruses cause serious diseases such as cervical cancer, and HPVs are not the only harmful biological agents. There are other highly contagious and devastating sexually transmitted diseases (STDs) that plague humanity. A partial list includes: herpes, chlamydia, gonorrhea, syphilis and acquired immunodeficiency syndrome (AIDS). Such diseases take a toll, not only on the individuals who become infected, but also on their families and on society at large. Neither Adam, nor Eve could have understood any of this biology, but given Adam's growing interest in sex, he was probably eager to do what Eve suggested.

Decisions bring consequences, and the garden episode was disastrous for everyone involved. Through her tryst with the serpent, Eve came to know the difference between good and evil, but what she did not know is that through her sexual contact with the seducer, she would thereafter carry new kinds of bacteria and viruses. The same was true for Adam.

Following their disobedience, the biological war that went on inside Adam and Eve grew increasingly fierce and complicated. Even so, their perfect DNA seems to have been powerful enough to prevent the full blown negative effects of the new biological invaders, at least for a time.

By today's standards, Adam lived a long time and had many children, but with each following generation, their DNA was increasingly diluted and corrupted. Neither of them continued in their state of ignorance; no longer were they naked and unashamed, and because of what they did, their offspring inherited a host of biological enemies. With her newly acquired wisdom, Eve knew what she had done was going to be problematic if God found out. Without advice from a mother or older sisters, what was she to do?

The Coverup

Well, what better way to assuage her own actions than by persuading Adam to do likewise? As the story continues, *she also gave some to her husband, who was with her, and he ate.* In the story I'm telling, Eve encouraged Adam to have sex with one of the other females in the Garden, the very thing God had told Adam not to do. Rather than heeding God's warning, Adam followed Eve's advice, and through their actions, Adam and Eve introduced sin and death into the Garden of Eden.

Since they are our ancestors, they dragged the rest of us into the quagmire. We can say Adam and Eve have given all of us some of the forbidden fruit. In this view, Eve's first sexual experience was with the serpent. He was a handsome, subtle creature who used his good looks and keen wit to entice Eve to disobey God. Adam and Eve exercised their free will to make choices that were not only catastrophic for them personally, but also for their children and for people today. Eve quickly realized she really messed up when she had sex with the serpent, and when Adam acted on Eve's advice, he was just as — if not more — guilty than Eve. After all, God told Adam not to eat the forbidden fruit, and this before Eve came along.

Obviously, sex is necessary for the continuation of the species, but as was true back then and is still true today, not every kind of sex is appropriate. Poor advice and poor results are handmaidens. Furthermore, poor actions on the part of parents can be devastating for their progeny.

Realizing what they had done, Adam and Eve made themselves some kind of clothing from fig leaves and hid from God. The Hebrew word for these garments is חֲגֹר (pronounced krag-ore). It's # H2290 in the Enhanced Strong's Lexicon. The KJV translates this Hebrew word as aprons and the NKJV translates it as coverings. The RSV translates as aprons, and the NRSV translates as loincloths. Collectively, these translations seem to suggest that what mattered to Adam and Eve was covering their genitals. They could not hide from God, and their loincloths of fig leaves were not sufficient to cover their sin. God did for Adam and Eve what they could not do for themselves when He fashioned different clothing for them.

The type of material God used for the new clothing is clear enough, but the style of the garment is somewhat ambiguous. The Hebrew word of interest is כְּתֹנֶת (pronounced, keth-o-neth). This is a different word than the one used to describe what Adam and Eve sewed together for themselves. This word is # 3801 in Strong's Enhanced Lexicon. The Abridged Brown-Driver-Briggs Hebrew-English Lexicon of the Old Testament (BDB) translates as '*tunic*, the principal ordinary garment worn next to the body.' Whatever the configuration of the garments, whether they were tunics, aprons or loincloths, they were intended to cover their nakedness. All translations agree God made the new clothing from animal skins.

Frankly, the present author's Sitz im Leben is such that loincloth seems to fit the bill. In this view, loincloth is another indication that their shame was the result of improper sexual behavior rather than poor menu choices. How so? Well, in order to eat anything, Eve and Adam would have passed the morsel into their mouths to be chewed and swallowed. If their sin was literally eating fruit from the wrong tree, and if they sought to cover their sin, which they clearly did, then it would seem more likely that they would have made veils to cover their faces, rather than loincloths to cover their genitals. Wearing loincloths of fig leaves, they hid themselves in the Garden, but God sought them out for an accounting.

Conjecture or Evidence

New ideas awaken (often loudly) the ever-present Sitz im Leben, and the conjectures presented herein are well outside the ruts of traditional interpretation of *the tree of the knowledge of good and evil*. So before we continue along this line, can we point to any evidence that might provide support for these novel views? Well, maybe.

The Genesis narrative on sin begins with an encounter between a woman and a serpent, and we have evidence of an ancient encounter between the two. The evidence comes in the form of a sculpture from what is known as the Gravettian Culture. The sculpture called *the Beauty and the Beast* is dated to about 23,000 years ago, and it depicts a woman and a serpent with their bodies intertwined. The two are connected at the back of their heads, at their shoulders and below the woman's waist. The sculpture is made of greenish-yellow stone gemologist called serpentine. To view the sculpture along with an unknown artist's motif of it, the reader may visit the following website and scroll down just a bit.

Parents are cautioned. The motif is sexually explicit and may be inappropriate for younger viewers.

http://anthropark.wz.cz/gravetta.htm

The caption on this page reads, in part: "The front view tells us that the first body is obviously a woman, whereas the second one is not a man, but a snake with a parted toothed mouth and apparently slanting eyes on both sides of the head. The belly of the snake bears notches that correspond to long transverse plates characteristic of bellies of the snakes. This apparently mythological motive of a snake and a woman appears in probably all mythologies of the world including Australia and Asia. The Old Testament told about Eve and the snake as well."

A scholarly treatment of these 'Venus figurines' was produced by Karen Diane Jennett.[55] In her paper, Jennett provides a thorough analysis of the Beauty and the Beast sculpture.

Is this small sculpture unequivocal evidence for the serpentine seduction of Eve I have suggested? No, it is not. However, it may be enough to steady those who are feeling a little shaky from reading these novel ideas. I'll say more about this Gravettian sculpture later.

The Reckoning

And they heard the sound of the Lord God walking in the garden in the cool of the day, and Adam and his wife hid themselves from the presence of the Lord God among the trees of the garden. Then the Lord God called to Adam and said to him, "Where are you?" So he said, "I heard Your voice in the garden, and I was afraid because I was naked; and I hid myself." And He said, "Who told you that you were naked? Have you eaten from the tree of which I commanded you that you should not eat?" Then the man said, "The woman whom You gave to be with me, she gave me of the tree, and I ate." And the Lord God said to the woman, "What is this you have done?" The woman said, "The serpent deceived me, and I ate." (Genesis 3:8-13)

Consequences for the Serpent

So the Lord God said to the serpent: "Because you have done this, You are cursed more than all cattle, And more than every beast of the field; On your belly you shall go, And you shall eat dust All the days of your life. And I will put enmity between you and the woman, And between your seed and her Seed; He shall bruise your head, And you shall bruise His heel." (Genesis 3:14-15)

During the Godly interview, the loquacious serpent goes silent. God does not ask for his version of the events. God accepted what Adam and Eve said on the matter, and judgment began with the handsome fellow who was more cunning that the other beasts.

Notice that God did not curse the serpent for what he said, but for what he did, as can be seen in the phrase, *"Because you have done this."* While it could be argued that what the serpent did was only what he said, this explanation ignores the plain meaning of the words. Talk is cheap. Actions speak more loudly, and some actions carry huge consequences. So what did the serpent do? Well, as discussed earlier, my conjecture is he seduced Eve.

Earlier I suggested the serpent was not a reptile. First, snakes don't talk. Second, they would not have the cognitive wherewithal to engage in conversation even if they could talk. Third, in response to what the serpent did, God cursed him to crawl on his belly. If the serpent was already crawling on his belly, as snakes do, then God's curse on him would have amounted to no curse at all, and God does not speak idle words.[56] Here is enough to show that the serpent was not a snake; he did not crawl on his belly when he seduced Eve.

What's more, in Eve's estimation, he was a delight to the eyes. In contemporary terms, the serpent was a good-looking, smooth-talking, bipedal scoundrel whose interest in Eve was sexual. This fully modern Homo sapiens fellow got what he wanted from Eve.

Later passages seem to suggest that Eve got pregnant from the encounter; more on this observation further down. God's righteous judgment fell on the guilty parties in ways that were commensurate with what they had done. After cursing the serpent, God told Eve how her life was going to change.

Consequences for Eve

To the woman he said, "I will greatly increase your pangs in childbearing; in pain you shall bring forth children, yet your desire shall be for your husband, and he shall rule over you." (Genesis 3:16)

The word, *curse*, is not found in this passage, but God declared He would increase Eve's pain in childbirth. Even so, Eve's desire would be for her husband. The implication is that if Eve had not done this thing, then her pain in giving birth would have been much less severe, perhaps negligible. How could this have been possible?

Well, some really big animals, such as grizzly bears, have babies that are tiny in comparison to the mother's size. Obviously, very small offspring would pass through the birth canal more easily than larger ones. If Eve had not conceived a hybrid child by having sex with the serpent, then maybe her babies would have been very much smaller than babies today; speculation, to be sure, but possible. However birth took place back then, mothers today will agree that giving birth is painful business, even with modern medicine.

In addition to her increased pain in child birth, God demoted Eve from her status of full equality with Adam. Thereafter, Adam would rule over Eve. We have encountered female atheists who despise this passage, claiming it to be little more than a chauvinistic ploy to oppress women. Such views are not surprising, for any passage of scripture that confronts human pride incurs the disdain of the narcissist, regardless of gender or religious convictions. Next, God pronounced judgment on Adam, and the passage just below does involve a *curse*.

Consequences for Adam

Then to Adam He said, "Because you have heeded the voice of your wife, and have eaten from the tree of which I commanded you, saying, 'You shall not eat of it': "Cursed is the ground for your sake; In toil you shall eat of it all the days of your life. Both thorns and thistles it shall bring forth for you, and you shall eat the herb of the field. In the sweat of your face you shall eat bread till you return to the ground, for out of it you were taken; for dust you are, and to dust you shall return." (Genesis 3:17-19)

Adam tried to excuse his behavior by blaming Eve for giving him bad advice, and even though God accepted this part of Adam's explanation, He did not exonerate the man. Human beings are social creatures, and each person occasionally receives advice from another. Sometimes, it's good advice; sometimes, it's bad. There is no sin in receiving either. What matters is what one does with the advice. As creatures with free will, we can ignore advice, or we can act upon it. God commanded Adam not to eat from the tree of knowledge of good and evil, but Adam gave heed to the voice of Eve. He chose to do that which God commanded him not to do. He became a sinner. Through Adam sin entered the world and death through sin.[57] This series of events is sometimes called *"the Fall."* Since we are their descendants, they took the rest of us along for the ride.

Instead of cursing Adam for heeding Eve's bad advice, God cursed the ground. Following the curse, the Earth brought forth thorns and thistles. As every farmer knows, cultivated soils produce more than only the kinds of crops the farmer plants. It also produces weeds (the thorns and thistles). It's a never ending battle, one Adam would fight for the rest of his life, and Adam did live a long time. So Adam went from light gardening duties to strenuous labor in order to get his food. We can say that as a result of their actions in the garden both, the man and the woman, would thereafter have hard labor, he in working the soil, and she in childbirth.

God told Adam that his body was made of dust, and to dust he would return. We know our bodies are made from atoms that are common in the dust of the Earth, and we are certain that following our death, the atoms of our bodies go off to do other things.

Old Man Adam

God had previously warned Adam that he would die in the day that he ate from the tree of the knowledge of good and evil — not that he might die at some later date, but that he would surely and immediately die. Eve persuaded Adam to eat the forbidden fruit, but instead of dying as God said would happen, Adam lived 930 years. So, how are we to understand this contradiction?

Was God mistaken about what would happen? Have we misunderstood the meaning of the phrase *"in the day?"* The Apostle Peter taught that *with the Lord one day is as a thousand years, and a thousand years as one day.*[58] Is this the problem? Or was Adam somehow pardoned? The idea that Eve or Adam knew more about the forbidden fruit than God did can be immediately dismissed. Even though Adam lived almost 1,000 years, which in St. Peter's understanding could be taken to mean *one day,* God's warning to Adam was one of immediate, physical death.

In order to answer this issue, it has been suggested that Adam died spiritually since his sin dramatically changed his original relationship with God. However, there is no mention of spiritual death in God's warning to Adam, so this interpretation seems a bit thin. The penalty for eating from the tree of knowledge of good and evil was physical death, but since Adam did not immediately die, there is something else going on in this story. What happened? Was the penalty set aside, something like a presidential pardon? Yes, it was, and here's how.

Also for Adam and his wife the Lord God made tunics of skin, and clothed them. (Genesis 3:21)

It is virtually impossible to overstate the importance of this short verse. With only 16 words, it tells what God did for Adam and Eve to save them from the death that was coming because of their sin. Through what God did it became possible for Adam to live 930 years instead of dying the same day he ate from the forbidden tree. Adam and Eve made aprons of fig leaves to cover their private skin, but fig leaves could not cover their public sin. More was needed to stave off the death which was coming. Even though they had no right to receive it, they got the help they needed from God. The help was a manifestation of God's grace.

Matters of Law

There are laws which govern the physical realm, and there are laws that govern the spiritual realm. Even though human beings have come to understand many of these laws, we had no part in formulating any of them. We did not establish the laws of nature, and we did not establish the laws that govern the spiritual realm. God did, and whether it suits our Sitz im Leben or not, death is the penalty for sin. Without exception all flesh shall die but not every soul. Some souls will enter heaven and other souls will not. Whether a soul goes to heaven or to hell depends on a phenomenon called *atonement*. Atonement can be understood as setting a soul back into proper relationship with God. Sin creates a gap that separates the sinner from God. Atonement closes the gap.

Another spiritual law describes the necessity of blood as the means of atonement. This law is clearly stated in the Book of Leviticus where we read the following:

For the life of the flesh is in the blood, and I have given it to you upon the altar to make atonement for your souls; for it is the blood that makes atonement for the soul. (Leviticus 17:11)

Life, blood and atonement are inextricably linked, and herein is the foundation of animal sacrifice that was an integral part of worship in the early Hebrew community. These blood sacrifices are seen by some Christians as a foreshadowing of the sacrificial death of Jesus. So why did God do this? Why did God shed an animal's blood for Adam and Eve? The one word answer is love.

Out of pure love God committed an act of divine grace when He took the skin from an animal to make tunics (aprons or loincloths) for Adam and Eve.[59] Adam did not ask God for these garments, and Adam certainly did not deserve them. Actually, Adam and Eve didn't even realize they needed different clothing. They thought they had their sin covered by the aprons of fig leaves they had made for themselves. Adam and Eve could not hide from God, and their aprons of fig leaves did not satisfy the spiritual law which required the shedding of blood to make atonement for their souls. Out of divine love, God shed the blood when He provided the coverings of animal skins Adam and Eve so desperately needed.

Grace is a foundational tenet of the Christian faith, and I'll say more about what God's grace means for humanity just below. For now, it is enough to understand that God intervened on behalf of Adam and Eve to make atonement for their souls. God's action allowed Adam to live on instead of dying the day he sinned. What else can we say about divine grace?

Simply defined, grace is "unmerited favor." The favor is from God, and humans have no merit to receive it. Grace flows from God's love, and Christians believe they are saved from sin by God's grace.[60]

Grace comes in different ways. In the case of Adam and Eve, grace was bestowed in the form of loincloths God made of animal skin. We are not told what kind of animal skin God used. I am entertained by the idea that God took the skin from a lamb. In this view, perhaps we might catch a glimpse of why John the Baptist, upon seeing Jesus coming for baptism in the Jordan River, said, *Behold! The Lamb of God who takes away the sin of the world!*[61]

What's more, we might glimpse the reason the Apostle John (not John the Baptist) referred to Jesus as *the Lamb slain from the foundation of the world.*[62] We know that Jesus was crucified about 2,000 years ago, and depending on how one understands the phrase, *the foundation of the world,* this could be very far into the past, or it could be taken to mean the foundation established in the Garden of Eden that fateful day. However one interprets the atonement for Adam's sin, we know that blood was shed on Adam's behalf because Adam did not immediately die. Adam lived more than nine centuries.

The Tree Of Life

There were two trees in the Garden of Eden that were different from all the others: *the tree of the knowledge of good and evil* and *the tree of life*. We have previously discussed the tree of knowledge as an allegory for inappropriate sex. Next, we come to *the tree of life*.

There are three ways one might interpret the tree of life. First, this was a tree with roots in the ground and fruit on the branches. Second, this story is a mythological tale with no basis in fact. Third, the tree of life is a metaphor for a cognitive being. In the first interpretation, the tree of life produced a fruit with beneficial health effects. Perhaps it contained some substance that kept the telomeres at the ends of genes from unraveling. Scientists are currently searching for chemical compounds that would extend life, and perhaps the tree of life produced what scientists are searching for today. In this view, the tree of life would be the botanical equivalent of the mythical fountain of youth.

In the second interpretation, the tree of life is part of a fantasy with no basis in fact, one fabricated to speak to questions that might arise when readers considered these texts. Perhaps it was meant to explain why people in each new generation were dying at younger ages.

The third possibility is that the tree of life is a metaphor for a cognitive being. We will proceed on the basis that the third interpretation is correct. In order to explore this conjecture and why the metaphor was used requires some background information.

Monotheism is the belief in one, and only one, God. Polytheism is the belief in two or more. Jews, Muslims and Christians are all monotheists, but they differ on how the one God is to be understood. Christians worship one God, and the God of Christians is the God of Abraham. Muslims believe they are the descendants of Abraham through his first born son, Ishmael. Jews believe they are descendants of Abraham through his second son, Isaac. In this view, these folks are cousins. When it comes to human understanding of God, the primary difference between Muslims, Jews and Christians is how Christians understand the One God.

For Christians, the One God exists in three persons: God the Father, God the Son and God the Holy Spirit. This is the Christian doctrine of the Holy Trinity, and attempts to explain this divine mystery tend to flummox the faithful and antagonize the atheist. Jews and Muslims reject the notion of the Holy Trinity. They consider it polytheism, but it is not. Christians believe in one, and only one, God.

God is ineffable; so how might we, with our finite understanding and limited toolkit, explain the Holy Trinity? Even though attempts to use human relationships to try and explain this heavenly mystery are inherently inadequate, we might point out that one person reading this sentence could be a father, a son and a brother. Another reader could be a mother, a sister and a daughter. Also inadequate, we can point out that water (H_2O) takes on one of three different forms depending upon the temperature. H_2O can be ice, liquid or steam, but the chemical nature of H_2O doesn't change. Only its form is different.

Although these examples are wholly insufficient to explain the divine mystery of the Triune God, they may be of some help to those who accuse Christians of polytheism. In fact, Christians worship one, and only one, God.

We are laying some foundations for how we might understand *the tree of life* as the Second Person of the Trinity, and His gracious actions toward Adam and Eve. We have no scientific evidence to support, or to refute, these claims. So we turn to the testimony of the Apostle John who opened his gospel account by speaking of the Second Person of Trinity as *the Word* when he wrote the following:

In the beginning was the Word, and the Word was with God, and the Word was God. He was in the beginning with God. All things came into being through him, and without him not one thing came into being. What has come into being in him was life, and the life was the light of all people. (John 1:1-4)

John tells his reader that from the beginning the *Word* was with God, and the *Word* was God. Furthermore, the *Word* was the source of everything, including life itself. This is not a polytheistic claim. Instead, it speaks of the first two Persons of the Holy Trinity, God the Father and God the Son. As the Second Person of the Trinity, the *Word* can appear in whatever form He pleases. For Christians, the *Word* became flesh in the person of Jesus of Nazareth. He did so expressly for the purpose of giving His life to make atonement for everyone who will accept His sacrifice. My conjecture is that the *tree of life* is a metaphor for the *Word* who is the Second Person of the Holy Trinity. This begs the question of why Moses used this metaphor instead of speaking plainly of the Second Person.

Clearly, the metaphor was not used because of indelicate subject matter, as was the case in the allegory of the *tree of knowledge of good and evil*. Instead, it may have been employed to protect the fledgling Hebrew community from themselves.

The Genesis narratives are for all people today, but the ex-slaves who came out of Egypt were the first audience, and if Moses had spoken directly about the Second Person of the Holy Trinity, it could have been a stumbling block that spawned a backward look to the religious practices of Egypt. The Sitz im Leben of the newly freed slaves was conditioned by the fact they had lived some 400 years in Egypt where polytheism was the order of the day. Their tendency to revert to the worship of the false gods of Egypt can be seen in the account of them demanding Aaron, the brother of Moses, forge a golden calf for them to worship while Moses was on Mt. Sinai.

Astonishingly, Aaron did what they wanted.[63] I suspect the power of peer pressure is rooted in the ancient and urgent need to be an accepted member of one's tribe. In Aaron's case, while Moses was on the mountain, his tribe consisted of the recently freed slaves. Whatever the reason, Aaron fashioned a golden calf as the ex-slaves demanded. Obviously, their Sitz im Leben was such that they were all too eager to worship false gods.

In the story I'm telling, the *Word*, the Second Person of the Holy Trinity, was in the Garden of Eden, and He was instrumental in the events that took place there. According to St. John, the *Word* is the author of life itself. What more fitting metaphor than *the tree of life* could one imagine to speak of the One who is the author of life?

The *Word* paid the death penalty that was coming to Adam, and in order to understand this saga, we remember that sin is the cause of the death of the soul. It is written, *"The soul who sins shall die."*[64] Sin was the problem for Adam, and it has continued to be the problem, not only for our ancestors, but also for people today.

As it is written, *"For there is no difference; for all have sinned and fall short of the glory of God."*[65]

God is Holy, and sin separates the sinner from God. A cornerstone of Christianity is that only God can forgive sin but will not do so without the shedding of blood. God does not break His spiritual laws. However, God can and does make atonement for the souls of sinners. But, and mark this well, God requires our participation in the process. When He does make atonement, the sinner is set back into right relationship with God. This setting back into right relationship is known as reconciliation. In order to be reconciled with God, atonement for sin must be made, and as previously noted, atonement requires the shedding of blood.

We find ourselves living in a time when talk of a blood sacrifice for sin is out of favor in some circles. In my humble opinion, God is not impressed with one's aversion to blood sacrifice. Not only so, but God may be offended since the blood is from the only begotten Son of God. Whether it suits our Sitz im Leben or not, atonement for the soul of people today only comes about through a very specific blood sacrifice. This central dogma of Christianity was set to music by Robert Lowry in his hymn titled *Nothing But the Blood*. Here's the first verse: "What can wash away my sins? Nothing but the blood of Jesus; What can make me whole again? Nothing but the blood of Jesus."

Christians believe the shed blood of Jesus Christ, and only His blood, can make atonement for their soul. Atonement not only sets the sinner back into right relationship with God in the present, it includes heaven as the final destination for the soul. However, if a person dies without having their sins forgiven, then their soul descends into hell where it will be destroyed. As discussed earlier, the idea the soul in hell lives on forever in torment is contrary to scripture. Human beings do not have a choice about how sin is forgiven. Sin can only and will only be forgiven through a sacrifice of blood, as it is written, *"Without shedding of blood there is no remission."*[66] Jesus gave His life to save ours.

No other sacrifice is sufficient, as Cain failed to recognize when he and Abel brought their sacrifices to God. More on this point later. So God told Adam that in the day he ate from the forbidden tree, he would surely die. Adam ate and he lived on. This means that Adam's sin was forgiven, and my conjecture is that *the Word*, as the *Tree of Life*, made the required blood sacrifice to make atonement for Adam and Eve. This act of God's grace was manifest in the tunics of animal skin that God made for Adam and Eve.

Although Adam did not immediately die, he and Eve faced other consequences for which they were ill prepared. It turned out that another intervention by *the Word* would be required as we'll see further down.

Even though sin can be forgiven through the shedding of blood, the earthly consequences of sin are not automatically set aside. Adam would come to understand this principle when God told him how his life was about to change and subsequently evicted him from the garden. Adam knew his actions warranted the sure and immediate death God had warned would follow his eating the forbidden fruit, but God did something extraordinary to save him from death. Through what God did for them, Adam and Eve came to understand that proper worship included a blood sacrifice.

Surely Adam and Eve passed this information along to their children. However, when the time for worship arrived, the sons of Eve brought different kinds of offerings. But I'm getting ahead of the story. We'll come back to this important point a bit further down.

God cursed the serpent to crawl on his belly, told Eve she would have pain in childbirth and be subject to her husband and told Adam he would have to work hard to get his food. Then God evicted Adam and Eve from the Garden of Eden.

Then the Lord God said, "Behold, the man has become like one of Us, to know good and evil. And now, lest he put out his hand and take also of the tree of life, and eat, and live forever"—therefore the Lord God sent him out of the garden of Eden to till the ground from which he was taken. So He drove out the man; and He placed cherubim at the east of the garden of Eden, and a flaming sword which turned every way, to guard the way to the tree of life. (Genesis 3:22-24)

Now here's an odd thing, at least at first glance. Before Adam ate the forbidden fruit, he did not know the difference between good and evil. How could this have been? How could the man with perfect DNA, the man who knew objects can have names, the man who knew how to farm — how could this man not have known the difference between good and evil?

Well, notions of good and evil come from two sources, Godly instructions and cultural traditions. Godly instruction is without error of any sort, but cultural traditions are flexible, and what is acceptable in one cultural setting may be outlandish in others. For example, genital mutilation of young girls in certain tribes in Africa is abominable to virtually all other cultures. And even though this practice has been declared illegal, not only is this procedure normative in some tribes today, it is required that young girls be circumcised in a manner that prevents them from having an orgasm. Apparently this ghastly procedure (certainly not from God) is believed to lessen the likelihood of adultery following marriage.

Whether we like it or not, God forbids certain kinds of sexual behavior. Homo sapiens have free will, but in those cases where Godly instruction is contrary to human tradition, the choice is clear, or at least it should be. Any cultural tradition must be judged according to what God has said on the matter. Adam certainly received Godly instruction, but he had no earthly parents, no tribe, no cultural norms describing what constitutes good and evil. Adam was cognitively superior to those of the evolutionary line of life, and God created Adam with the capacity to exercise free will. Even so, the texts state that only after he disobeyed God did he come to know the difference between good and evil.

In order for Adam to *become* (anything), he must have been different before the events that evoked the change. So before the forbidden fruit episode, Adam's choice was to heed what God said and behave accordingly, or to heed Eve's advice and to act on what she said. We know what Adam did. Through his disobedience, Adam's innocence came to an end. So the Lord declared Adam had come to know the difference between good and evil, exactly what the serpent told Eve would happen if she would do what he wanted. This ability to know the difference between good and evil is an attribute God specifically associates with the image of God when He said, "Behold, the man has become like one of Us, to know good and evil."

Earlier, we argued the males and females mentioned in the first chapter of Genesis were not Adam and Eve. A corollary which supports this claim can be seen in that Adam came to know the difference between good and evil only after he ate the forbidden fruit. This lends credence to the conjecture that the people in the first chapter of Genesis were not Adam and Eve. Otherwise, Adam would have already known the difference between good and evil before the forbidden fruit episode.

So two abilities, the capacity for dominion over other kinds of life and the ability to know the difference between good and evil, are both in view when considering what it means to be created in God's image. There is a noticeable difference in how these areas of knowledge came to be. God gave the anatomically modern Homo sapiens the capacity for dominion over other animals and the capacity to know the difference between good and evil. In contrast, Adam and Eve attained the knowledge of good and evil through their sin. The sin, even though it was forgiven, still carried earthly consequences. God relieved Adam from his gardening duties and evicted the couple from the Garden of Eden. Even though God made atonement for Adam's sin, Adam's life circumstances were forever changed. He left the Garden of Eden knowing the difference between good and evil.

The story continues with Adam and Eve in their new circumstances outside the Garden.

Life On The Outside

Concerning Sex

The fourth chapter of Genesis opens with Adam and Eve consummating their husband and wife relationship outside the Garden and the subsequent birth of two children. For convenience, here is the text from the New King James Version of the Bible:

Now Adam knew Eve his wife, and she conceived and bore Cain, and said, "I have acquired a man from the Lord." Then she bore again, this time his brother Abel. (Genesis 4:1-2)

Remembering Maimonides observations, we point out that the phrase, *Adam knew Eve,* means the two of them had sexual intercourse. This is the first mention of Adam and Eve having sex with each other. Notice that the text is structured in a manner which makes Cain the son of Adam. This is deeply ingrained church dogma, and the power of the Sitz im Leben will make any other interpretation seem untenable, at least at first. Eve clearly states that she received help from the Lord in the delivery of her first child. We'll come back to this very important comment by Eve. It is perfectly clear that Cain and Abel were Eve's first two children, but it is not as clear that these two births were separated by months as would be expected in the usual course of human reproduction. In addition, it is possible these two births were not the result of two separate acts of intercourse between Adam and Eve.

It has been suggested that Cain and Abel were twins. The Reverend John Calvin believed they were, and Jeff A. Brenner, an authority on the ancient Hebrew language, agrees. Brenner writes,

"Notice that there is only one conception, but two births. The Hebrew word we translate as "again" is *asaph*. It means to add to. In this case the birthing of Abel was added to the birthing of Cain. Cain and Abel were twins."[67] Other commentators disagree, and some are silent on the question.

DNA certainly determines our physiology and cognitive abilities. If DNA orchestrates human behavior as much as we are beginning to think it does, then we may be able to gain some insight by looking at how the boys behaved. If Cain and Abel were twins, as Calvin and Brenner suggested, then in the story I'm telling, they had different fathers.

Is it possible for a woman to give birth to twins with different fathers? Well, yes, it is possible, and scientists call it *superfecundation*. Fecund, from the Latin word, *fecundus*, means abundant production or very fertile. Superfecundation in this case speaks to the production of more than one ovum during a single menstrual cycle. Twins with different fathers can be produced when a super-fecund woman has sex with different males during her time of super fertility. This is not speculation. According to a well-documented paternity case, a woman gave birth to twins with different fathers. Genetic tests confirmed that one of the woman's ovum was fertilized by one man, and her second ovum was fertilized by a different man.[68] So, according to science, a woman giving birth to twins with different fathers is a real possibility.

My conjecture is the serpentine seduction inside the Garden resulted in Eve's first pregnancy, and her coupling with Adam outside the Garden produced her second. In this view, Eve was carrying twins. The suggestion that Cain and Abel had different fathers hinges not only on the glaring difference between the behavior of the two boys, but also, as Brenner has pointed out, on the Hebrew word which the versions above translate as *again*. The essence of this word is "add to, continue or repeat that which precedes it." In this view, the texts suggest, as Calvin and Brenner have pointed out, Cain and Abel were twins, but when they wrote, these two commentators overlooked the possibility that the boys had different fathers.

If things worked back then rather like they do today, then about nine months after the serpentine seduction, Eve went into labor, and if anyone ever needed help, she certainly did. How could a woman with no belly button know what to do with the umbilical cord? Without the help of a physician, a mother or a midwife, how could Eve have given birth to her first son? Eve needed help, and she got help from the *Lord*. The Lord in this passage certainly was not Adam. Instead, the Lord who helped Eve with her birthing experience is the same Lord who earlier provided the couple with loincloths made of animal skins. I have suggested *the tree of life* is a metaphor for the Second Person of the Holy Trinity. He was with God in the beginning, and He was in the Garden of Eden. Inside the Garden of Eden, He made atonement for Adam and Eve. Outside the Garden, He helped Eve with the birth of her son.

As discussed above, there are some who accuse Christians of worshipping more than one God because of the doctrine of the Holy Trinity. Christians worship one, and only one, God — not three. Even though we cannot establish from the text why Moses spoke of the *Word* in metaphor, since the writing of the Apostle John, we have been in position to speak plainly about the *Logos*, the *Word of God* who became flesh as Jesus of Nazareth for the explicit purpose of giving His life (shedding His blood) to make atonement for sin.

To tidy up a bit, my conjecture is Eve was already pregnant with the serpent's son when she and Adam were evicted. When she and Adam had sex outside the Garden, Eve conceived her second pregnancy — Adam's son Abel. When her time came, she gave birth to twin boys. She gave birth to Cain first, and she declared she got the help she needed from the Lord. *The Lord* is the Second Person of the Holy Trinity, and for the reasons described earlier, He is portrayed as *the Tree of Life* in the story. Following the birth of Cain she again gave birth, this time to Abel. Considering Cain's paternal DNA, perhaps we can glimpse a possible answer to the question of how Eve's elder son could have murdered his younger brother.

Human beings did not invent sex, but we do have opinions on what kinds of sex are appropriate. Because we have free will, we can act upon our own desires or upon the advice of others. Or we can act upon what God has said on the issue. These choices are not always mutually exclusive, but when they are, then we do things our way, our advisor's way or God's way. Eve chose to act on the serpent's advice, and Adam chose to act on Eve's advice. They both ignored God's instruction, and all of humanity has been paying the price ever since.

Concerning Worship

In the story I'm telling, *the serpent* is a metaphor for one of the anatomically modern Homo sapiens who had made the changeover to fully modern behavior through his interactions with Adam. This fellow seduced Eve, and she became pregnant with Cain. In this view, Cain got half his DNA from the serpent and the other half from Eve. The DNA of the serpent included the genetic mutations accumulated in many previous generations. The serpent's ancestors were creatures of instinct whose actions were not guided by a sense of morality or ethics. They did what was required in order to survive and to take advantage of breeding opportunities that came their way. Eve's DNA was perfect, but the DNA of the serpent was a mess, and Cain got a dose of both.

As discussed earlier, through the animal skin clothing episode in the Garden of Eden, Adam and Eve came to understand the necessity of a blood sacrifice as part of worship. They would have passed this information along to their children. Unfortunately, as parents know, providing children with accurate information does not necessarily mean the children will make the right decisions. The evolutionary line of hominids were survivalists who did whatever was necessary, no matter how ghastly, to stay alive and to get what they wanted. At times, they went so far as to cannibalize other hominids. Even so, individuals must have had some sense of what they could do without suffering retribution from others in their tribe. Whatever their standards were, their children did not come prewired for compliance to those standards. They had to learn what constitutes acceptable behavior.

However, simply knowing the standards does not guarantee compliance. As criminals do today, some of the early hominids probably ignored the cultural standards in their efforts to get what they wanted. We are, at least in part, descendants of these evolutionary survivalists, and elements of their selfish ethos is etched into the DNA of modern people. In this view, the human predilection for war and the ongoing need for police and prisons may be somewhat easier to understand.

Christians might say we behave badly due to our sinful nature, and it seems to me that our sinful nature is deeply rooted in the corrupt DNA we inherited from our evolutionary ancestors. As discussed in Part II, we are beginning to understand the role of DNA, not only on our physical being, but also on our behavior. There is no doubt Cain and Abel, even though they were twins of Eve, behaved very differently. In due time, Cain became a farmer, and Abel became a shepherd. Just as Eve disregarded what Adam told her about the forbidden fruit, Cain ignored what his parents told him about the necessity of blood sacrifice when making an offering to God. Here is the relevant text.

And in the process of time it came to pass that Cain brought an offering of the fruit of the ground to the Lord. Abel also brought of the firstborn of his flock and of their fat. And the Lord respected Abel and his offering, but He did not respect Cain and his offering. And Cain was very angry, and his countenance fell. So the Lord said to Cain, "Why are you angry? And why has your countenance fallen? If you do well, will you not be accepted? And if you do not do well, sin lies at the door. And its desire is for you, but you should rule over it." (Genesis 4:3-7)

The two sons of Eve each brought an offering as part of their worship. The problem for Cain was he chose to bring an offering of grain which did not meet the blood requirement. Abel, on the other hand, brought an animal from his flock which did. God did not respect Cain or his offering, but God did respect Abel and his offering. Because they had free will, Eve's sons could have brought, and in fact, did bring, whatever they wanted as an offering to God. Abel brought his animal offering as a matter of faith. He believed what his parents told him, and he acted accordingly. Apparently, Cain did not believe blood was a necessary part of offering, and he also acted accordingly. God chastised Cain for his grain offering, but assured him if he acted properly in the future, he would be accepted. Cain not only ignored his parent's instructions, he also ignored God's instructions. Rather than listen to God and bring a proper offering the next time, Cain murdered his brother. Here is the relevant text.

Now Cain talked with Abel his brother; and it came to pass, when they were in the field, that Cain rose up against Abel his brother and killed him. (Genesis 4:8)

We have no record of the conversation between the boys, but we know Cain murdered his younger brother. Poor Abel; the only thing he did was bring a proper offering as part of his worship of God. Murder is abhorrent in any situation, but when the victim is a family member, the dastardly act becomes even more contemptible. What could have caused Cain to do these things?

We know that genes regulate our morphology and physiology, and a rapidly growing body of scientific evidence is showing that the part of DNA called the epigenome influences not only our bodies, but also our behavior. As discussed in Part II, the epigenome regulates gene expression, and we have discovered that the epigenome can be modified by traumatic experiences. In what has been called the "nature versus nurture" debate, not only does nurture (the environment) of a child affect his or her behavior, but astonishingly, the child's behavior is also influenced by what happened to the parents and grandparents before the child was born. So not only does the epigenome we inherit from our ancestors regulate gene expression, it also plays a very important role in our behavior.

Professor Isabelle Mansuy at the University of Zurich and her team have demonstrated that childhood trauma has a lifelong influence on blood composition, and that these changes are passed along to the following generation. She reported, "These findings are extremely important for medicine, as this is the first time that a connection between early trauma and metabolic disorders in descendants is characterized."[69] Traumatic experiences modify the epigenome, which in turn, modifies the expression of genes, which in turn affects behavior.

This discovery is so important that it bears repeating. Trauma affects the behavior, not only of the person who experiences the trauma, but also the behavior of their descendants. Perhaps this is why the scripture mentions the iniquity of the fathers passing onto their children, and the children's children, unto the third and fourth generation. We can barely imagine the circumstances of our ancestors. Not only were they under daily threat of being eaten by predators, but, at times, by other hominids. So environmental considerations, especially the more extreme episodes, affect not only those individuals personally involved in the episodes, but also the following generations.

In comparison to Adam, the serpent's DNA was a genetic mess. Not only did it carry the mutations accumulated over thousands of generations, but also the epigenetic anomalies generated through the traumatic experiences of his ancestors.

If the serpent really was Cain's father, then perhaps the murder is a bit more understandable. In this view, we can say that the murder of Abel is the clearest indicator that Cain's paternal DNA was from the serpent. By today's standards, Cain's father was an immoral brute, and from a genetic perspective, it seems Cain was a chip off the old block. But what Eve did was not Cain's fault. And if Cain had been open to God's instructions, he might have avoided further problems. Following the murder of Abel, God questioned Cain. Here is the relevant text.

Then the Lord said to Cain, "Where is your brother Abel?" "I don't know," he replied. "Am I my brother's keeper?" The Lord said, "What have you done? Listen! Your brother's blood cries out to me from the ground. Now you are under a curse and driven from the ground, which opened its mouth to receive your brother's blood from your hand. When you work the ground, it will no longer yield its crops for you. You will be a restless wanderer on the earth." (Genesis 4:-12)

God cursed Cain, and the curse altered Cain's relationship with the Earth. This is a mystery; thereafter the Earth would no longer yield its strength to Cain. Furthermore, Cain would be a fugitive and vagabond, a restless wanderer. There was no court of appeal, but Cain complained to God that his punishment was more than he could bear.

And Cain said to the Lord, "My punishment is greater than I can bear! Surely You have driven me out this day from the face of the ground; I shall be hidden from Your face; I shall be a fugitive and a vagabond on the earth, and it will happen that anyone who finds me will kill me." (Genesis 4:13-14)

One of the most informative parts of this conversation is what Cain said would happen to him. He told God anyone who saw him would kill him. Adam and Eve would have many children over the ensuing centuries, but Cain did not fear retribution from yet to be born siblings. He feared those already living in the area. Notice that God did not contradict Cain's assessment of the situation. God did not say anything like, "What you fear cannot happen for there are no other people out there." Why did God not correct Cain on this matter? Because Cain was right; there were other people out there. Cain knew it, and, obviously, God knew it.

These other people were the anatomically modern Homo sapiens who came out of Africa, some of whom had made the changeover to fully modern Homo sapiens through their interactions with Adam. Cain's biological father (the serpent) was such a man. As noted above, these hominids carried the DNA of their ancestors who, for many generations, had lived under the constant threat of being eaten by predators and even by other hominids. Generations of extreme violence colored their epigenome. These were dangerous folks, and Cain knew it.

In response to Cain's analysis of what would happen when he encountered the others, God, in an amazing act of grace, put a mark on Cain warning anyone who met Cain not to kill him. The nature of this mark is unknown. Following Cain's conversation with God, Cain went into the land of Nod where he continued farming, but with less productive results. Permanent settlements cannot be sustained without agricultural enterprise to feed the residents; hunter-gatherers do not construct cities. Cain's farming enterprise allowed him to build a city, and he took a wife from among those Homo sapiens already living in the area. The two of them had a son they named Enoch. And due to genetic mixing, Enoch's DNA was even more corrupt than was Cain's.

Cain was sentenced to a difficult life outside the garden, and his descendants were destined to have their DNA increasingly corrupted through genetic mixing with the evolutionary kinds of hominids. Ironically, some of the people Cain feared would kill him became part of Cain's family of farmers.

In order to understand what came to pass a bit further along in the Genesis narrative, it is important to understand that the female descendants of this hybridized line of folks are called the *daughters of men*. The male descendants of Adam and Eve, those who got their Y chromosome from Adam and their X chromosome from Eve, are called the *sons of God*. The perfect DNA of Adam is long gone, and although it might be entertaining to speculate that we might someday recover it through genetic engineering using CRISPER/cas 9 or some other as yet to be discovered technology, it seems, at least to me, virtually impossible that we shall ever do so. People today carry some DNA from Adam, some from Neanderthals and some from the serpent. If epigenetic imprinting of our DNA really does influence our behavior, then perhaps we might glimpse the reason why human history is littered with disgusting behavior, including what seems to me the most repugnant of them all — war.

Sex is an extremely powerful part of the human story. Obviously, it is necessary for the continuation of the species, and carried out according to Godly rules, things go pretty well, even in light of the faulty DNA we each carry. However, sexual practices that are contrary to Godly rules continue to wreak havoc on those who ignore the rules, just as it did for Adam and Eve.

Life Grows Shorter

The serpent in the Garden of Eden was one of the evolutionary line of Homo sapiens who had made the changeover to fully modern behavior through his interactions with Adam. This fellow was *cunning*, and Eve thought he was *a delight to the eyes*. He seduced Eve and became the biological father of Cain. This view is supported in the New Testament where we read the following: *For this is the message that you heard from the beginning, that we should love one another, not as Cain who was of the wicked one and murdered his brother.*[70] Cain was not Adam's son. He *was of the wicked one*, and his paternal DNA is on display in the murder of his own half-brother. The serpent's DNA carried not only the genetic mutations accumulated over previous generations, but also the various bacteria and viruses inherited from his ancestors. We have identified many diseases that are linked to genetic mutations. The serpent was handsome. He was cunning, and he wanted to have sex with Eve. And in the story I'm telling, he did.

The Genesis narrative suggests that Adam would have lived a very long time if he had not eaten the forbidden fruit, but whatever the original plan was and regardless of the events that changed the plan, Adam did live a long time, at least by today's standards. We know physiology is orchestrated by DNA, and it seems Adam's longevity was due to the pristine DNA God gave in him to begin with. In this view, even though he was exposed to a battery of new biological invaders through his sexual contact with the evolutionary line of hominids, his immune system held up pretty well for nearly a thousand years. Even so, after 930 years, Adam died, and his body returned to the dust from which he was created.

With each new case of genetic mixing that took place outside the Garden of Eden, the perfect molecule of DNA God gave Adam was diluted with the mutated DNA of the evolutionary line of hominids. This genetic corruption continued when Cain took a wife from the evolutionary line of hominids and the two of them had children. This pattern of dilution and corruption of the perfect molecule of DNA shortened the lifespans of the generations that followed. To be clear, my conjecture is that the perfect molecule of DNA of Adam was repeatedly diluted and increasingly corrupted through genetic mixing that took place in his progeny. This genetic mixing weakened the ability of the owners to defend against biological enemies.

The genetic mixing also reduced the ability of the body to repair itself. The evidence shows that the DNA of people living today includes many mutations. So either God gave Adam faulty DNA to begin with, or the perfect DNA that Adam had at the beginning was corrupted over time. The pristine DNA molecule is no more, and as a result, nobody today lives hundreds of years.

Life Grows Complicated

The genetic mixing that God warned against also resulted in unusual children. Some of these children were characterized by extreme wickedness, some by violence, some by extraordinary size and some by all three. These hybrids are called n*ephilim*. These bizarre hominids were not the result of interbreeding between angels and women. They were the result of the various mixings of the perfect and powerful DNA of Adam with the mutated DNA of the evolutionary line of hominids. Here is the text.

Now it came to pass, when men began to multiply on the face of the earth, and daughters were born to them, that the sons of God saw the daughters of men, that they were beautiful; and they took wives for themselves of all whom they chose. (Genesis 6:1-2)

We'll come back to this segment a bit further down. For the present we turn to the question of whether or not God is omniscient.

Bad Things Happen

When some tragedy overtakes a respected member of the community, someone will usually ask a question such as, "Why does God let bad things happen to good people?" This was precisely the case when a tornado killed my nephew and his wife. Keith Walls was an outstanding young man, one of the finest I've ever known. He and wife, Donna, had a 6-month old son who survived the tornado with barely a scratch. As had happened before in similar circumstances, I was asked the question at their funeral. Basically, the question can be generalized in this fashion: Where is God in all this?

Those who ask what God is like are often told that God is omniscient, which means that God knows all things, at all times and in all circumstances. Even though this claim is deeply engrained in church doctrine, we can ask whether or not it is accurate. If God truly is omniscient, then some passages of scripture become very difficult to understand. Science is understandably silent on this issue. So what is the scriptural evidence on whether or not God is omniscient?

Well, after Adam and Eve ate the forbidden fruit, they hid from God.[71] Scripture says God called out to Adam with this question, "*Where are you?*" Some believe God posed the question in order to make Adam aware of his situation. This interpretation ignores the substance of God's question, for unless we rob the words of their ordinary meaning, the question indicates that God did not know where Adam was hiding. What's more, Adam certainly knew that he had messed up. Otherwise, why hide from God?

After learning their hiding place, God asked Adam why he was hiding and subsequently, who told him he was naked. Then God asked Adam if he had eaten from the forbidden tree. After Adam's answer, God asked Eve what she had done. These questions suggest that God did not know everything, but asked Adam and Eve in order to learn the facts of the matter.

The same logic applies to God's interview of Cain after Cain murdered his younger brother. As noted earlier, God asked Cain, *"Where is Abel your brother?"* Cain answered, *"I do not know. Am I my brother's keeper?"* God then asked Cain another question, *"What have you done? The voice of your brother's blood cries out to Me from the Ground."* These questions can hardly be taken as intended to make Cain aware of his transgression. Cain's answers make it very clear he already understood what he had done.

Another example can be seen in Genesis 22 where we read the account of God testing Abraham by commanding him to sacrifice his son, Isaac. As the story unfolds, Abraham was about to sacrifice Isaac when God stopped him. The angel of the Lord said to Abraham, *"... for now I know that you fear God, since you have not withheld your son, your only son, from Me."* The key phrase being *"now I know"* which indicates God did not know beforehand what Abram would do.

Although God can come to know whatever He wants, the above examples suggest that He does not necessarily know everything in advance. The reason God does not know everything in advance is due to His amazing gift of "free will" to Homo sapiens.

If the above examples are not sufficient, Genesis 6:6 should settle the question of God's omniscience. In this passage we read the following. *"And the Lord was sorry that He had made man on the earth, and He was grieved in His heart."* [72]

If God is omniscient, if He knows the end of everything from the beginning, and if God was sorry for what He had done, would He not have done things differently beforehand? Of course, He would have. The scriptural evidence shows that God does not know all things at all times. Such is the power of free will. It not only makes the future pliable, but also mediates the tedium we imagine would result from omniscience. Still, God can come to know whatever He chooses.

Also notice the previous text (the one that describes the Lord's sorrow that He had made man on the earth) is not about Adam and Eve. God was not sorry that He formed Adam from the dust and cloned Eve from Adam's body. Instead, His sorrow was connected with the evolutionary line of hominids. My best guess is God was not sorry that He created them in the first place, but sorry that He blessed them. The blessing enabled them to become fully modern through their interactions with Adam, and in this manner, they gained "dominion" over the other forms of evolutionary life. When they made the changeover to fully modern behavior, they became *more cunning* than the other beasts the Lord had created. With their heightened cognitive abilities, including the ability to exercise their free will, things went haywire.

God can certainly know whatever He pleases, but God is not omniscient, per se. Not every tornado, hurricane, earthquake, fire or other disaster is on God's radar. Bad stuff happens, and it is incorrect to say that God causes all of it. On the other hand, can God bring famine or some other natural disaster on a land? Of course. But not every disaster is from God.

A Family Tree

Sons of God and Daughters of Men

God created Adam and Eve for each other, not for the other hominids who occasionally meandered through the Garden of Eden. However, Eve and Adam chose to disobey God and not surprisingly, things went horribly wrong. We have no way of knowing Eve's biological age when her encounter with the serpent took place, but in the story I'm telling, she was of child bearing age since she became pregnant with Cain. Lacking any scientific evidence to support or refute this conjecture, we turn first to scripture and second, to a bit of circumstantial evidence which might bear on these issues.

Without doubt, the genealogy of Adam (his toladah) begins in the fifth chapter of Genesis, and neither of Eve's first two sons are included in the list of Adam's descendants. Why not? Abel is not listed because Cain murdered him before Abel had children. However, Cain did have children, but he is not listed in the genealogy of Adam, either. Again, why not?

The omission of Cain from the genealogy of Adam is not because he murdered his brother. Sons are not left out of their father's genealogy, not even for ghastly behavior. For example, King David committed adultery with Bathsheba, the wife of Uriah. She became pregnant, and in his attempt to coverup what he had done, David arranged for Uriah to be killed in battle. He murdered Uriah by proxy.[73] Yet David is a central character in Hebrew scripture. The point is sons are not left out of genealogies, not even for egregious behavior. What's more, as noted earlier, the New Testament clearly states that Cain *was of the evil one.*[74] Cain is not listed in the genealogy of Adam because he was not Adam's son.

Sculptures & Skeletons

Adultery is understood as consensual sex between a married person and a person who is not his or her spouse. God explicitly forbids this behavior.[75] Adultery was forbidden in the Hebrew community, so much so that if discovered, the man and woman faced capital punishment.[76] Any woman, especially one who was pregnant by someone other than her husband, was sure to suffer the scorn of society, including public stoning in ancient Israel. So secrecy was extremely important to those involved in adultery.

Obviously, men do not get pregnant. They do not get a swollen belly that heralds their every step, but a woman can only hide her pregnancy for a few months. In the days before abortion, a woman who became pregnant through adultery was left with few options. She had to claim she was raped, or convince her husband that the child she was carrying was his. With the advent of birth control and abortion, the threat of one's adultery becoming public knowledge has lessened, but the possibility of the adultery becoming public knowledge never goes completely away.

Even though public scorn for adultery may be on the wane in America, since God has condemned sex outside marriage, only among infidels might adultery ever become acceptable behavior. No one who commits adultery can keep their secret from God. Adultery really is egregious behavior, and it produces what we can call a skeleton in the closet. In this metaphor, the skeleton is knowledge of the adultery, and the skeleton is in the closet because the owner wants to keep the adultery secret.

Skeletons can be pesky things. Even when the owner manages to keep his or her skeleton in the closet for years, the fear of the skeleton getting out never totally goes away. If the skeleton does get out of the closet, the disclosure can be catastrophic, as a former President of the United States learned. (Many readers will remember the saga of the infamous "blue dress.") Rumors of a handsome fellow (not Adam) seducing our great, great, great… grandmother, Eve, would qualify as this sort of skeleton.

Adam and Eve lived with the consequences of their sin all their lives. They tried to hide what they had done from God, which they could not, and they probably wanted to hide it from their children. However, adultery rarely remains secret for very long, especially when the perpetrator is someone famous. Neither did Eve's sexual encounter with the serpent remain secret. Instead, it may have become part of ancient stories whispered around campfires.

Whispers don't fossilize, but stone sculptures can last thousands of years. Earlier, we mentioned a bit of circumstantial evidence that may have bearing these matters. Perhaps that which was once whispered around the campfires took on a more permanent form as an artifact associated with Gravettian culture. I have not included all the graphics pertaining to this sculpture for they may be too explicit for younger readers. In order to more easily follow the arguments below, the reader who has not already done so is again encouraged to view the sculpture and associated materials by visiting this website:

http://anthropark.wz.cz/gravetta.htm

Gravettian

The Gravettian culture dates to about 28,000 — 19,000 BCE. The name comes from 'la Gravette,' the archaeological site in southwest France where artifacts from the culture were first discovered. Debate continues on when the first fully modern Homo sapiens arrived on the stage of history, but without doubt, the Gravettians were fully modern. Dozens of their sculptures called "Venus figurines" have been found across Europe, from France to Siberia.

The artifact of interest is a small sculpture which has been dated to about 23,000 BCE, and it seems to suggest Eve's skeleton took on a more permanent form after it got out of the closet. An unknown Gravettian artisan took a piece of pale, green-yellow, serpentine stone and shaped it into a sculpture called "the beauty and the beast" (*La Belle et la Bete*, in French). It's no more than coincidence that this sculpture is crafted from a material that modern gemologists call serpentine. This small sculpture depicts a woman (the beauty) and some sort of animal (the beast) connected at the backs of their heads, at their shoulders and below the woman's waist. The animal in the sculpture clearly has scales, but it also has arms and is in a standing position. If this sculpture really is a depiction of the encounter between Eve and the serpent, this is evidence that the seducer was physically very different from snakes today.

However, if the viewer of the sculpture is unaware that the serpent in the Garden of Eden was different from snakes today, he or she may have difficulty identifying the animal in this sculpture as a serpent if for no other reason than, as everyone knows, serpents don't stand upright. As expected, the ability to see the animal as a serpent (or not) will depend on the Sitz im Leben of the viewer.

In the usual course of events, that which ends in consensual sex begins with the eyes. It might start with a flirtatious glance from a woman who has spotted what she considers to be a really good looking guy, or maybe with a jaw-dropping stare at the most beautiful woman a man has ever seen. The particulars don't really matter, but it begins with the eyes. Obviously, the information acquired through the eyes must be processed in the brain. If the parties find each other physically attractive, and if they are interested in the possibilities, then the episode usually continues with conversation. If the conversation is satisfactory to both parties, some sort of touching will follow. The kind of touching doesn't matter; it could be something as mundane as a slight touch of the hands, or something more intense such as dancing together. If the encounter has progressed far enough, the touching might become more pronounced. The point is physical contact of some form or the other will take place before the sexual intercourse.

If this Gravettian sculpture actually does speak to the serpentine seduction of Eve, dating this figurine places a lower limit on when Eve's tryst with the serpent took place. How so? Well, artists don't create sculptures meant to depict some historical event before the event takes place. So, if this beauty and the beast figurine really does depict Eve's sexual encounter with the serpent, then the seduction took place before about 23,000 BCE which is the estimated date the figurine was crafted. Below is a photograph of the beauty and the beast sculpture; courtesy of donsmaps.com.

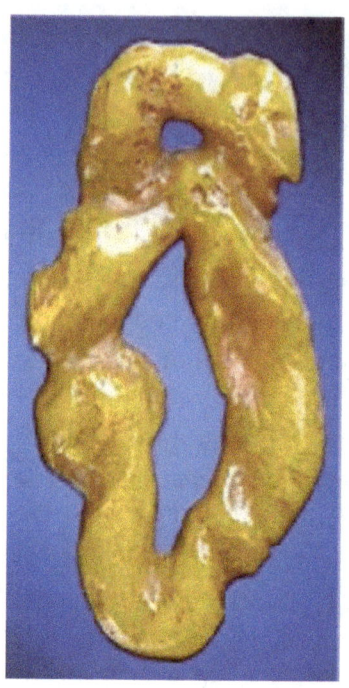

Photo: © Harry Foster

As mentioned earlier, that which ends in consensual sex begins with limited information collected by the eyes and processed in the brain, information the person interprets something like, "She is an attractive woman." Or "He's an attractive man." When it comes to potential sexual partners, rarely does one choose to pursue someone they consider unattractive, at least if there are other options available. Hence the crude witticism sometimes heard in night clubs and saloons, "All the girls get prettier at closing time." In the quest for sexual partners, if one person finds another attractive, if they are seen as a delight to the eyes, conversation follows. These flirtations are mental engagement in so much that the participants must be of like minds in order for the episode to go forward. In Eve's judgment the serpent was a *delight to the eyes*. The Garden episode continued with witty conversation. Through his lies, the serpent managed to engage Eve on a mental level. Perhaps the artist meant to reflect this mental engagement by the connection of their heads. This is speculation. Some sort of touching came later, and the touch convinced Eve the fruit actually was *good for food* since she didn't die as she had said would happen. Maybe the artist meant to reflect the touching by the connection of their shoulders. The serpent seduced Eve, and perhaps the artist meant to reflect the sex act by the connection of their bodies below the woman's waist.

We cannot know for certain what the artist had in mind, but whatever it was, it involved a woman and some kind of serpent.

On Eve's advice, Adam disobeyed God by having sex with one of the anatomically modern Homo sapiens in the Garden of Eden. Humanity has been paying the price ever since.

Shoulders Of Giants

As mentioned in Part I, Sir Issac Newton said, "If I have seen further than others, it is by standing on the shoulders of giants." In the following, we find ourselves also standing on the shoulders of giants, but not in the positive way that Newton meant. We have unequivocal evidence which shows genetic mixing between different kinds of hominids took place in the relatively recent past. This mixing produced hybrid offspring who carried various, never seen before, combinations of genetic material, and some of these were explosive combinations. According to the Biblical account, not only were some of these hybrids very evil, they were very large. Here is the relevant text from the NKJV.

Now it came to pass, when men began to multiply on the face of the earth, and daughters were born to them, that the sons of God saw the daughters of men, that they were beautiful; and they took wives for themselves of all whom they chose. And the Lord said, "My Spirit shall not strive with man forever, for he is indeed flesh; yet his days shall be one hundred and twenty years." There were giants on the earth in those days, and also afterward, when the sons of God came in to the daughters of men and they bore children to them. Those were the mighty men who were of old, men of renown. Then the Lord saw that the wickedness of man was great in the earth, and that every intent of the thoughts of his heart was only evil continually. And the Lord was sorry that He had made man on the earth, and He was grieved in His heart. So the Lord said, "I will destroy man whom I have created from the face of the earth, both man and beast, creeping things and birds of the air, for I am sorry that I have made them" But Noah found grace in the eyes of the LORD. (Genesis 6:1-8)

One truth claim in this passage is *the sons of God* and *the daughters of men* produced hybrid children. As already discussed, science claims and evidence confirms that different kinds of hominids produced hybrid children in the relatively recent past. These claims from scripture and science are complementary, not contradictory.

Prior to our understanding that different kinds of hominids had hybrid children, theologians proposed various explanations to explain who the *sons of God* and the *daughters of men* were, including the idea that the sons of God were angels. They were not. Adam is called *the son of God* in the New Testament.[77] The sons of God were the male descendants of Adam and Eve, males with no evolutionary DNA within their genome. They carried the pristine, powerful DNA of their parents. The daughters of men were females whose genome contained DNA from the evolutionary line of hominids. For examples, Seth, as the child of Adam and Eve, was one of the sons of God, and the woman Cain married was one the evolutionary line of Homo sapiens. She was one of the daughters of men.

Remembering that Eve considered the forbidden fruit (the serpent) a delight to the eyes, notice that the beauty of the daughters of men was a consideration in the sons of God taking the daughters of men as wives. Beauty really is in the eye of the beholder, and although physical beauty is desirable, it can be deceitful.

One truth claim in the above passage is when the sons of God mixed with the daughters of men, some of their children were much larger than average, so much larger that they were called *giants*. As is well understood, breeding different kinds of animals can produce offspring that are different from either parent. Herdsmen have been cross breeding animals for many decades in efforts to produce more desirable (often larger) offspring.

In addition to the Genesis account of these giants, the authors of the books of Deuteronomy and Joshua identify groups who were descendants of these giants and who were still alive after Adam and Eve were long gone.[78] At least two groups of large people lived in the land of Canaan during the time of the conquest by the Hebrews. They are called the *Anakim* and the *Emim*. A King named Og was the last of these very large people. His bed was nine cubits long and four cubits wide. The ancient cubit was about 18 inches. This bed which was 13.5 feet long was known to be in the city of Rabbah.

In addition, we have extra-biblical literature that confirms there really were giant hominids in the past. The secular historian, Flavius Josephus, wrote in his *Antiquities of the Jews* that the bones of these giants were still on display in the city of Hebron (Israel) during his lifetime.[79]

Some believe that the Biblical accounts of giants are no more than fables. It can be difficult for some modern investigators to believe there really were giants in the land in those times, but to dismiss both the claims of scripture and the testimony of Josephus seems, at least to me, a step too far. The genomes of the sons of God and of the daughters of men were similar enough to produce offspring, but the genomes were different enough that interbreeding, at least at times, produced children of extraordinary size. It is also possible that some of the hybrid children were cognitively different from either of their parents, and some may have had other non-reported physical anomalies we have not imagined. This is speculation, but the biblical texts are clear that another consequence of this genetic mixing was the increase in evil. We just don't yet know enough to sort it all out, but the truth claim of scripture and the written testimony of Flavius Josephus agree. There were giants in the land in those days.

The interbreeding between the *sons of God* and the *daughters of men* produced unusually large offspring, but why did this mixing result in the increase in evil behavior? Stated differently, why were the giants so evil? My conjecture is the genetic mixing affected not only their size, but also their thinking and consequentially, their behavior. In addition, they likely were able to impose their will on others due to their superior size and strength. The daughters of men carried evolutionary DNA, and the sons of God carried DNA that God gave Adam. Apparently, the combination of the potent DNA of Adam and the corrupt DNA of the evolutionary line of hominids was a powerful but disastrous combination.

As noted earlier, we have discovered that environmental conditions can impact the epigenome in ways which we are just now beginning to understand. Further research in this area by Dr. Isabelle Mansuy at the University of Zurich has shown that trauma influences both the victim and their progeny. Dr. Mansuy said, "Our findings demonstrate that early trauma influences both mental and physical health in adulthoods and across generations, which can be seen in factors like lipid metabolism and glucose levels."[80]

The small tribes of anatomically modern Homo sapiens lived with virtually constant trauma. Predators chased them down and ate them. The faster individuals almost certainly heard the screams of slower members of their tribe being dragged to the ground and torn apart. If geneticists are correct in their belief that traumatic experiences affect not only the operation of the epigenome of the one traumatized, but also that such changes are inherited by their progeny, then the children born from the mixing of the sons of God with the daughters of men were also affected by the traumatic experiences of their progenitors. In this view, the saying, "The apple doesn't fall far from the tree." gains some credence. Not only so, but since the early hominids were also our ancestors, we carry some of the epigenetic scars inherited from those who came before us.

As parents understand all too well, young children are selfish little buggers. Children come hardwired to think any action that gets them what they want is permissible; they have no concept of right and wrong or good and evil. Such values must be taught by parents, and most, but not all, parents do an adequate job of it. However, judging from the number of prisons in America, it is evident that some individuals never move beyond their penchant for predatory practices. For such as these, the kinds of toys change, but the selfish behavior does not. Children clash over cookies, pacifiers, dolls, toy cars and such. Unsupervised two year olds are likely to injure each other as they struggle over toys.

Adults clash over more grandiose toys — especially land and money.[81] History is littered with accounts of those who started wars in their efforts to take land and money from others. Alexander the Great, Julius Caesar, Genghis Kahn, Joseph Stalin, Adolph Hitler and Mao Tse-tung come to mind. This desire to control others and their toys has not gone away. Vladimir Putin has made the list. Human beings are still selfish.

The descendants of Adam and Eve interbred with the evolutionary line of hominids. This genetic mixing plunged the world into great wickedness, and the LORD was sorry He had made man, and He was grieved in His heart.[82] In the story I'm telling, God was not sorry He made Adam. He was sorry he made those of the evolutionary line mentioned in the first chapter of Genesis. My best guess is God was not sorry that He created the early forms of animal life in the first place, but sorry He enhanced the genome of some of them when He blessed them to give them dominion over the others. Here we have unequivocal evidence that God allowed life to proceed with a great deal of autonomy. If God had controlled every aspect of evolutionary history to some foregone conclusion, then He would not have been sorry that what He wanted to happen actually did happen.

God lit the fires of life, but allowed them to burn without His constant control. Here is evidence of free will at work, and in this case the exercise of free will resulted in circumstances that grieved God. God's solution for this societal mess was a flood to destroy man and beast.

We read in verse eight of Genesis 6, *"Noah found grace in the eyes of the LORD."* As stated earlier, grace is unmerited favor; the favor is from God, and those who receive it have no merit to do so. In this instance, grace came in the form of a Godly plan and instructions on how to carry out the plan. Explaining the plan, God told Noah that destruction was coming, and what he had to do in order to save himself, his family and other kinds of life. God told Noah how to build the boat, and God's instructions to Noah did not result in a leaky vessel. Those who went on board were saved from the flood. If Noah had ignored God's plan or decided to build a different kind of boat, then he would have perished along with all the others. The same observation applies to people today. We are not under the threat of destruction by a great flood, but we are under the threat of destruction of our soul in the fires of hell. The Old Testament declares the situation in simple, clear and unequivocal terms:

"Behold, all souls are Mine; The soul of the father as well as the soul of the son is Mine; The soul who sins shall die." (Ezekiel 18:4)

This passage is not about death of the body. Every person will surely die, and will do so in the relatively near future. Clearly, this passage is about death of the soul. As discussed earlier in Part III (Concerning The Soul) eternal torment of the soul is not part of God's plan. Souls of the unforgiven are not tortured for eternity. They descend into hell where they are obliterated.

Sin is the problem, and it is a universal problem because all have sinned. Everyone needs to be saved, and anyone can be saved. The soul of the saved will live on forever in heaven, but being saved does not happen automatically. God told Noah what to do to be saved, and God has told us what we must do to be saved. As was true for Noah is true for people today. Those who get on board with God's plan need not fear; they will be saved. Their soul will not die. Those who reject God's plan will perish.

Modern folks need tools. Carpenters need saws and hammers, electricians need wire cutters, warriors need weapons, sailors need ships, doctors need stethoscopes, astronomers need telescopes and so on. We not only need the right tools for the job, we need to know how to use them. We don't drive nails with a stethoscope. In the following we describe the tools God has given us and make clear how to use them. Doing so guarantees our soul will "land safe on that happy shore" when we die. When all else fails, read the instructions!

The preceding materials are a mixture of facts, conjecture and speculation. I have been careful to distinguish between them. The following material does not contain speculation or conjecture. Instead, the remainder of this book contains the facts the reader needs in order to make the most important decision of his or her life.

The Decision

> I want to know one thing, — the way to heaven: how to land safe on that happy shore. God himself has condescended to teach the way. He hath written it down in a book. O, give me that book! At any price, give me the book of God! I have it: here is knowledge enough for me. Let me be a man of one book.
>
> — The Reverend John Wesley

As previously discussed, there are laws which govern the operation of the physical realm and laws which govern the spiritual realm. In addition, there are man-made laws meant to regulate human behaviors. We'll call these natural, spiritual and man-made laws respectively. The laws that govern the physical realm were in existence long before hominids walked on the Earth. Obviously, none of us had any part in formulating the natural laws. Neither did we have any part in formulating the spiritual laws. The natural and the spiritual laws are what they are, whether we like them or not. Obviously, human beings are responsible for the content of the man-made laws.

It is not possible to break natural laws, but human beings can, and frequently do, break spiritual laws and man-made laws. Breaking a spiritual law is sin. Breaking a man-made law is crime.

There is an inescapable penalty for breaking spiritual law, and there are potential penalties for breaking man-made law. No one can avoid the penalty for breaking spiritual law, but some criminals manage to evade punishment for their crimes against man-made laws. Penalty and consequence are not necessarily the same things. For instance, incarceration is often the penalty for a serious crime known as a felony, but a consequence is difficulty in finding employment following release from prison. Some employers will not hire a convicted felon. There must be penalties for violating man-made laws in so much that laws without penalties are little more than suggestions. Prisons are populated with people who thought they could commit crime without being caught and punished.

Some crimes are heinous, but others are minor infractions. Hence, there are different penalties for different types of crimes. Capital punishment (the death penalty) is reserved for those who commit the most egregious crimes, but parking violations are usually punished with a small monetary fine. In contrast, there is one penalty for sin. As noted earlier, "the soul who sins shall die."

In some cases the different kinds of laws overlap, but not in others. For example, spiritual law forbids murder and so does man-made law. This means murder is both a sin and a crime. On the other hand, the natural laws have nothing to say about murder. The cosmos cares nothing about whether one person murders another. The human experience takes place in an environment shaped by natural, spiritual and man-made laws. One might obey the spiritual laws most of the time, but breaking a single law, just once, makes one a sinner. For illustration, the person who does not murder or steal is innocent with regard to these two laws, but if the same person breaks the Sabbath or bears false witness against someone, that person is a sinner. One drop of dye changes the color of all the water in the glass. Likewise, one sin stains the soul. The one who transgresses God's laws, any of the laws even once, is a sinner.

Sin is more than the proverbial fly in the ointment; it's the elephant in the room. The Old Testament Hebrew word for sin is חָטָא (pronounced khaw-**taw**). It's a verb which means to "miss" (a goal or way), to go wrong.[83] The New Testament Greek word for sin is ἁμαρτία (pronounced, ham-ar-**tee**-ah).[84] This word can be associated with archery, particularly with shooting an arrow which misses the target. We can say that sin is missing the target (described by some spiritual law) God has said we should aim for and hit.

Very simply, sin is any violation of God's spiritual law. Sin is universal. It started with Eve, continued with Adam, and it has remained unabated ever since. Sin not only damages our relationship with others but also our relationship with God. God is holy, and sin separates the sinner from God in the present, and unless the sin is forgiven before death, the separation becomes permanent. It's not as though the separated soul exists for eternity, for it does not. Instead, as discussed earlier, the soul that is separated from God when the sinner dies will be destroyed. This is the final penalty for sin.

When Christians speak of salvation, they are referring to the forgiveness of sin which releases them from penalty of death hanging over their soul. Thankfully, sin can be forgiven before the Grim Reaper comes calling. And when the sin is forgiven, the sinner is reconciled with God, which means he or she is set back into proper relationship with God in the here and now, and the proper relationship continues after death. The person whose sin is forgiven is said to be "saved." Forgiveness of sin is the only antidote for the elephant in the room.

What else do we know about salvation? Well, salvation includes two components: God's *grace* and human *faith*. We'll discuss grace first and then faith. In his epistle to the church at Ephesus the Apostle Paul spoke of salvation this way.

"For by grace you have been saved through faith, and that not of yourselves; it is the gift of God, not of works, lest anyone should boast." (Ephesians 2:8)

Grace

Grace can be defined as unmerited favor. The favor is from God, and those who receive God's grace have done nothing to deserve it. Stated differently, grace cannot be earned. Grace is a gift — the ultimate expression of God's love. We cannot emphasize this point enough: God's grace is available to everyone, despite the gravity of their sin(s).

Everyone needs to be saved. But not everyone will be saved, for God does not override the free will He has given us. Very sadly, some refuse God's grace, and those who do remain under the law of sin and death.

Stages of Grace

God's grace comes in three stages: *prevenient, justifying* and *sanctifying*. Receipt of the second and third stages of grace depends on how the recipient responds to the first stage. It is not necessary for the one who wants to be saved to hold a Ph.D. in theology or to be highly fluent in the particulars of these terms, but in order to be saved, a certain response to the prevenient grace is absolutely necessary. First, God gifts the sinner with prevenient grace.

Prevenient Grace

Salvation can be understood as a process. The first step in the process begins as God awakens the person to the reality of their condition. For instance, when Adam and Eve sinned, they did not approach God seeking forgiveness. To the contrary, they hid from God. But God sought them out. God always initiates the process of salvation. Likewise, Noah didn't approach God in anticipation of the flood. God approached Noah. The point is that God initiates salvation. Speaking of these matters, Jesus said, *"No one can come to me unless drawn by the Father who sent me; and I will raise that person up on the last day."*[85] God initiates the process of salvation as He gives prevenient grace.

So, how does this come about? As the time for His departure drew near, Jesus explained it to his disciples as follows.

> *"Nevertheless I tell you the truth. It is to your advantage that I go away; for if I do not go away, the Helper will not come to you; but if I depart, I will send Him to you. And when He has come, He will convict the world of sin, and of righteousness, and of judgment..."* (John 16:7–8)

The Helper is the Holy Spirit, and the Spirit comes to initiate the process of salvation. As Jesus explained in the passage above, the Spirit will "convict the world of sin, of righteousness and judgment." Stated a bit differently, the Spirit convinces and convicts the person of his or her sin. Absent the work of the Spirit, the sinner remains mostly, if not entirely, oblivious to their condition.

For illustration, children do not come into the world with any sense of good and evil or right and wrong. Parents must instill these values in their children. If they fail to do so, the children are likely to do physical harm to their playmates in order to maintain control over the toys. Hitting, biting, kicking and pushing are common behaviors in small children, at least until parents make the children understand that such behavior is not acceptable. Parents awakening their children to these issues can be likened to the Helper (the Holy Spirit) awakening adults to their sinful condition.

Through prevenient grace, God begins the process of salvation. Some accept this grace. Others do not. Prevenient grace enables, but does not ensure acceptance of the gift of salvation. Those who reject God's grace have various stories to explain and console themselves.

For example, some might go so far as to claim they are strangers to sin. They may or may not acknowledge that sin is real. Those who do acknowledge the reality of sin may claim they have never worshipped a false idol, never taken God's name in vain, never murdered and never stolen anything. Such as these rely on their own behavior (good works), if not to earn God's approval then at least to convince themselves they are good people who have no need for a personal savior. I suspect they would be less willing to claim they never broke the Sabbath.[86] Furthermore, when those who were relying on their good behavior to earn God's approval asked Jesus about the commands of God, He answered:

"You shall love the Lord your God with all your heart, and with all your soul, and with all your mind. This is the first and great commandment. And the second is like it: You shall love your neighbor as yourself. On these two commandments hang all the law and the prophets." (Matthew 22:37-40)

(Every time Jesus quoted "scripture," He was quoting the Old Testament. In this answer just above Jesus quoted Deuteronomy 6:4 and Leviticus 19:18.)

Few of those who rely on their own good behavior would claim to love their neighbor as they love themselves. Christians know, as the Apostle Paul wrote in the Book of Romans:

"For there is no difference; for all have sinned and fall short of the glory of God." (Romans 3:23)

We have all sinned. All men know in their heart that they have chapters of their life they do not want published. God knows our sin, but until we are willing to acknowledge our sin, we cannot confess our sin, and without confession, we remain under the penalty for our sin. As it is written in the New Testament Book of First John:

"If we say that we have no sin, we deceive ourselves, and the truth is not in us. If we confess our sins, He is faithful and just to forgive us our sins and to cleanse us from all unrighteousness. If we say that we have not sinned, we make Him a liar, and His word is not in us." (I John 1:8-10)

God's prevenient grace is meant to awaken us to our sinful condition, and the Holy Spirit is the agent of God's grace. Jesus' promise of the Holy Spirit came to fruition when the Spirit descended on a large group of ethnically diverse people gathered in Jerusalem. The Christian Church was born that day. It's called Pentecost. The Apostle Peter explained what was happening that day, and upon hearing Peter's explanation, some asked Peter what they should do. Peter answered:

"Repent and be baptized every one of you in the name of Jesus Christ for the remission of sins, and you shall receive the gift of the Holy Spirit." (Acts 2:36-39)

The Greek word translated as "repent" is μετανοέω (pronounced met·an·o·**eh**·o).[87] Repent means "change of mind." Since one's mind directs one's path, I find it helpful to think of repentance as a "change of direction." For illustration, let Miami, Florida, represent heaven, and let Seattle, Washington, represent hell. (These cities are chosen for their fame and distance from each other; nothing more.) The traveler who wishes to arrive in Miami cannot keep driving toward Seattle and expect to end up in Miami.

The traveler must repent, must change directions. Does the change of direction instantaneously put the traveler in Miami? No. The change only points the traveler in the right direction. To change directions, to head for the desired destination — in this case, heaven — this is repentance.

However, even after repentance, we still have a problem, which we can explain as follows. Suppose a person decided he will never sin again and somehow managed to pull it off. The sins committed previously are still part of the record. In order to make it into heaven, the previous sins must be expunged from the record. And only God can do this. So God's prevenient grace awakens us to our condition, and if we are willing, we confess our sin and we repent. These are the first steps in the process of salvation.

The one who has confessed and repented will want to take the next step. They will want to be baptized in the name of Jesus Christ for the remission of sin, as the Apostle Peter explained in the passage above. Remission of sin wipes the penalty of death off the books, if you will. Regardless of how hard one tries, no one can live a life free of sin. But God has provided a way out of the mess we have created. Better yet, God has provided the *only* way.

The way (the only way) is Jesus Christ, the Son of God. Jesus was without sin, and He gave His life (His blood) to pay the sin debt of the world. This is God's plan of salvation. And there is *no other plan*, as clearly stated in the Book of Acts.

"There is salvation in no one else, for there is no other name under heaven given among mortals by which we must be saved." (Acts 4:12)

In addition, Jesus clearly stated that He was the only way to the Father when He said, *"I am the way, and the truth, and the life. No one comes to the Father except through me."* (John 14:6)

This is why we are baptized in the name of Jesus Christ. God offers salvation to everyone, but God does not force salvation on even one. Contrary to claims that Christianity is exclusionary, God does not want anyone to perish. He wants everyone to "land safe on that happy shore." Even so, God has not provided smorgasbord of salvation options. Jesus is the only way to heaven.

The next stage in the process of salvation also comes through God's grace — through His justifying grace.

Justifying Grace

The spiritual laws of God do not change, and, as noted earlier, *without shedding of blood there is no remission.* Remission can be understood as payment of a debt. The debt is incurred by sin, and the shedding of blood is the only currency that can pay this debt. So what has God done in order to satisfy what the law requires? The answer to this question is at the heart of the Christian religion. Sin and death are inextricably linked, and death is synonymous with the shedding of blood. Whether it suits one's Sitz im Leben or not, there is no way around any of this; the spiritual laws of God are immutable.

God told Adam not to eat the fruit from the forbidden tree, but Adam did it anyway. Whether the reader accepts the forbidden sex conjecture described earlier or prefers the traditional interpretation that Adam ate fruit from the wrong tree, either way, Adam sinned, and he did so under the penalty of death. Yet, Adam lived over nine centuries. How was this possible?

As the spiritual law requires, Adam's debt to sin was paid through the shedding of blood. The blood was from the animal whose skin God took to make the garments for Adam and Eve. If we are to be saved, where's the shed blood that pays for our sins?

As mentioned earlier, Christians believe that Jesus gave His life (His blood) to pay the sin debt of the world. In what is arguably the most famous passage in the New Testament, Jesus told the story as follows:

"For God so loved the world that He gave His only begotten Son, that whoever believes in Him should not perish but have everlasting life. For God did not send His Son into the world to condemn the world, but that the world through Him might be saved." (John 3:16-17)

From its beginning, the church has taught that this passage speaks to the sacrificial death of Jesus Christ who was falsely accused by the Jewish authorities, vilified, scourged and crucified by the Roman military. Isaiah was a prominent Old Testament prophet. In chapter 53 of the book which bears his name he wrote prophecies about these matters, and he did so hundreds of years before Jesus was born. Furthermore, Psalm 22 describes in some detail what took place on that fateful day. This psalm was written centuries before Jesus was born. It is through Jesus' sacrifice, through the shedding of His blood, that the penalty for our sin is paid.

The sacrificial death of the only begotten Son is the ultimate example of God's love for humanity. Those who accept His death as atonement for their sin will "land safe on that happy shore" as Reverend John Wesley put it. Those who reject God's offer will experience a dramatically different outcome.

All this is a manifestation of God's grace which God offers to all people — not only to a select few, but to everyone. God's promise is that when we confess our sin, repent and ask Jesus to forgive us, He does. At this juncture the death penalty hanging over souls is removed. Our debt to sin is paid by Jesus, the Lamb of God, as John the baptist called Him. God's grace is received through faith. We are saved by grace, through faith. We have discussed the first two kinds of grace: prevenient and justifying grace. Now we come to the third kind of God's grace.

Sanctifying Grace

Even after one is saved, there is more, and the more is once again made possible by God's grace. This next phase in the Christian walk is empowered by God's *sanctifying* grace. What do we know about this type of grace?

Sanctifying grace is meant to lead those who have been saved into the fullness of the Christian life. Sometimes this part of the Christian experience is called *going on to perfection*. The author of the Book of Hebrews explained as follows.

"Therefore let us go on toward perfection, leaving behind the basic teaching about Christ, and not laying again the foundation: repentance from dead works and faith toward God, instruction about baptisms, laying on of hands, resurrection of the dead, and eternal judgment." (Hebrews 6:1-2, NRSV)

After laying out his impeccable Hebrew religious credentials, the Apostle Paul, writing to the church at Philippi put it this way:

"But what things were gain to me, these I have counted loss for Christ. Yet indeed I also count all things loss for the excellence of the knowledge of Christ Jesus my Lord, for whom I have suffered the loss of all things, and count them as rubbish, that I may gain Christ and be found in Him, not having my own righteousness, which is from the law, but that which is through faith in Christ, the righteousness which is from God by faith; that I may know Him and the power of His resurrection, and the fellowship of His

sufferings, being conformed to His death, if, by any means, I may attain to the resurrection from the dead." (Philippians 3:7-11)

Christianity recognizes a number of people who have lived exemplary lives devoted to God. Mother Teresa comes to mind, and we might argue that such saints reached perfection before their time on Earth came to an end. Even so, most Christians would consider themselves a work in progress, as did the Apostle Paul in the passage above. This progress is empowered by God's sanctifying grace. Few would claim to have achieved perfection.

Wherever one is positioned along the road of life: One who denies the existence of God; one who believes in God, but remains unaware of their sin before God; one under conviction of sin, but still unwilling to confess and repent; one having confessed and repented, but not yet baptized; one baptized and going on to perfection — no progress from one position to another is possible aside from God's grace. If we are saved at all, we are saved by grace through faith.

However, faith is not what some suppose it to be. Some think faith is only what we believe, and what we believe is clearly part of faith. However, faith is more than only what we believe.

Faith

Depending on the context, faith can be a noun or a verb. Students of the English language are taught that nouns are words which denote people, places or things. And verbs are words which denote actions. Faith as a noun can speak to the set of beliefs held by a particular group of people: the Christian Faith, the Jewish Faith or the Muslim Faith, for examples. Faith as a verb denotes the actions (works) people undertake as a result of their beliefs. The confusions between faith (as a noun), faith (as a verb) and beliefs (in general) has a long history.

The New Testament comes down to us in the Greek language. The Greek words translated into English as "faith" and "believe" are from the group which scholars call the "Pist Group." Examples from the pist group include: *pisteuo, pístis, pistos* and pistóō.[88] Few, if any other words, are more thoroughly discussed in scholarly literature. And thankfully so. Faith is a richer concept than some suppose.

Some think faith is no more than what they believe. This view is not totally mistaken — just incomplete. What one believes is clearly a large part of faith. However, faith is more than *only* what one believes.

When we say we believe something, we mean we are in mental agreement with the set of propositions that pertain to the something in which we believe. For examples: We believe habanero peppers are hot to the taste. We believe we can drive safely through the intersection when the traffic light is green. We believe many, many things. Even as a verb, faith includes components of belief because one's actions (faith) will be determined by what one believes. However, faith is more than *only* mental agreement with a set of propositions.

Salvation is the ultimate goal. It occurs when grace combines with faith. It's not that grace needs anything additional for God's grace is totally sufficient. And God's grace is universally available.

But grace saves only those who evince faith. Again, what is this faith that leads to salvation?

Various words from the P*ist* group are used to denote: belief, trust, reliance and faith. The verb is often used for believing God's word, for believing the prophets and especially for believing in Jesus and His words and deeds. P*istis* is primarily faith in Jesus Christ. Faith in Christ means believing in His identity, His teachings and in His death and resurrection. Perhaps an analogy from biology can help.

The reader will remember from our discussions in Part II that a cell is a container with smaller containers inside. Those smaller containers are called *organelles*. We can think about faith in terms of cellular structure. In this analogy "faith" corresponds to the entire cell with two organelles inside: "belief" and "works." Furthermore, just as the cell contains organelles, the organelles also contain elements.

Continuing the analogy, the elements inside the belief organelle can be characterized as the *set of statements* which function to clarify religious convictions. These statements usually begin with the phrase, "I believe." It's this set of statements that fill the "belief" organelle. At this point the question becomes: What *exactly* do Christians believe?

Over the centuries, the church has formulated various documents to delineate and clarify exactly what Christians believe. Such formal statements of beliefs are known as "creeds." For example, *The Apostles Creed* is an ancient creed recited aloud and in unison in many churches during worship. It's also sometimes called an *Affirmation of Faith*. (The formatting of the creed below corresponds to the lyrical recitation by the worshippers.)

I believe in God, the Father almighty,
Creator of heaven and earth,
and in Jesus Christ, his only Son, our Lord,
who was conceived by the Holy Spirit,

> *born of the Virgin Mary,*
> *suffered under Pontius Pilate,*
> *was crucified, died and was buried;*
> *he descended into hell;*
> *on the third day he rose again from the dead;*
> *he ascended into heaven,*
> *and is seated at the right hand of God the Father almighty;*
> *from there he will come to judge the living and the dead.*
> *I believe in the Holy Spirit,*
> *the holy catholic Church,*
> *the communion of saints,*
> *the forgiveness of sins,*
> *the resurrection of the body,*
> *and life everlasting. Amen.*

Obviously, the first affirmation is belief in God. Absent this foundation, none of the other claims make any sense. The creed continues with affirmations of what God has done — especially in Jesus Christ. This creed has stood the test of time because it represents what so many Christians actually do believe.

The second organelle inside the cell of faith is "works." Works are those things people do as a result of what they believe. Tensions between early church leaders centered largely on the relationship between *faith* and *works*.

The Apostles Peter and Paul had jarring differences on what was required of a Christian.[89] One particularly gnarly dispute arose on the question of whether or not gentile (non-Jewish) Christians had to undergo circumcision. Unfortunately, other sorts of disputes on faith and works still surface from time to time.

Perhaps we can explicate these issues by describing two extreme positions. One the one end of the debate were those who took the position that Christians had to keep *many* (if not all) the Jewish laws. And there were hundreds of these laws —

circumcision being central. On the other end of the debate were those who took the position that *none* of the laws applied to them. They could do *anything* they wanted. Both of these positions are in error.

New Christians certainly do not have to undergo circumcision. Neither are Christians free to do *anything* they want. Christians are free from Jewish ceremonial laws, but not free from moral laws. For example, nobody can be taken seriously who claims that idolatry or murder or adultery are permissible for Christians. Various shades of these extreme positions still roam the halls of some churches today.

Nevertheless, one's actions cannot be separated from what one believes. The two are woven together as intimately as the double helix of DNA. The Apostle James spoke to the mistaken idea that faith and works are fully unrelated. We'll come to his remarks on this topic a bit further down.

In the above analogy of faith as a biological cell which contains other bodies, the other bodies are "beliefs" and "works." Rightly understood, faith is a dynamic combination of what we believe and what we do because of what we believe.

We can say the faith necessary for salvation is the faith that believes in someone and trusts in something. Christian faith believes in God, believes in Jesus Christ as the Son of God and trusts that Jesus gave His life to atone for their sin. They believe Jesus was crucified, dead and buried. They also believe Jesus rose from the dead on the third day. Christians believe these proceedings are manifestations of God's grace. These are key elements in the plan of salvation. These elements of faith are matters of belief and trust and actions shaped accordingly.

This plan of salvation (the gospel) is seriously good news for those who would like to "land safe on that happy shore." It is good news because God has done for us what we do not deserve and what we cannot do for ourselves.

A devout Pharisee named Saul persecuted the early church — at least until his conversion. His conversion was dramatic and profound. Following his conversion, he became known as Paul. Paul spent the remainder of his life preaching Christ, and Him crucified and raised from the dead. Paul wrote most of the New Testament. In his view, faith is in one person, and Jesus of Nazareth is that person. Drawing from the Old Testament Book of Habakkuk 2:4c, Paul explained as follows:

For I am not ashamed of the gospel; it is the power of God for salvation to everyone who has faith, to the Jew first and also to the Greek. For in it the righteousness of God is revealed through faith for faith; as it is written, 'The one who is righteous will live by faith'. (Romans 1:16)

As discussed above, Christian faith includes the set of actions we take (works) because of what we believe. For a simple illustration of how action is conditioned by what we believe, suppose one believes fire has broken out in the theater where he is seated. Undoubtedly, the person will take action. If possible, he will leave the theater. If this person does not believe the theater is on fire, he will likely choose to remain seated and watch the performance.

Likewise, if a person is diagnosed with cancer and told by her physician a certain treatment will eradicate the disease, she will almost certainly do something to get the treatment. What one believes does affect what one does. As regards salvation of the soul, it's not as though the works save. It's God's grace that saves. But faith without works is not the kind of faith that leads to salvation. The New Testament Book of James speaks to the relationship between faith and works:

> *What does it profit, my brethren, if someone says he has faith but does not have works? Can faith save him? If a brother or sister is naked and destitute of daily food, and one of you says to them, "Depart in peace, be warmed and filled," but you do not give them the things which are needed for the body, what does it profit? Thus also faith by itself, if it does not have works, is dead. But someone will say, "You have faith, and I have works." Show me your faith without your works, and I will show you my faith by my works. You believe that there is one God. You do well. Even the demons believe—and tremble! But do you want to know, O foolish man, that faith without works is dead?* (James 2:14-19)

In the biblical sense, "works" are those actions driven by religious convictions. Faith includes belief in God, trust in God, trust in His plan of salvation and human actions (works) shaped accordingly. When one accepts God's prevenient grace and thereby enters into the process of salvation, his or her religious convictions change. This change inevitably leads to a different set of actions, to a different set of works. Sins which had been pleasurable will become very distasteful, and those actions which held little or no interest will become delightful.

For example, those who never seriously read the Bible will want to study it. Those who rarely (if ever) went to church will become regular attendees. Those who stole, told lies and broke the Sabbath will stop doing such things, not under compulsion, but because these activities become repugnant. Instead, they will begin doing (with pleasure) those things pleasing to God, things called "good works." Even so, the good works do not save our soul. Very importantly, good works are *consequential* rather than *causative*.

To tidy up a bit, Christian faith includes the belief that Jesus is the Son of God, that He was born of the virgin Mary, that He suffered under Pontius Pilate, that He was crucified, dead and buried. Jesus gave His life to make atonement for sin. Faith includes the belief that Jesus rose from the dead on the third day, that He was seen by many witnesses and that He ascended into Heaven where He makes intercession for those who accept Him as Lord and Savior.

The consummate example of God's expectation for Christian actions can been seen in Jesus' instructions to his disciples. It's called "The Great Commission."

"All authority has been given to Me in heaven and on earth. Go therefore and make disciples of all the nations, baptizing them in the name of the Father and of the Son and of the Holy Spirit, teaching them to observe all things that I have commanded you; and lo, I am with you always, even to the end of the age." (Matthew 28:18-20)

The person baptized is said to be "born again." He or she is saved. When this life is over their soul will "land safe on that happy shore."

The above explanation of faith may seem too complicated. And in a certain sense, it is. However, it is appropriate because some misunderstand faith and mistake their newly discovered freedom in Christ as a license for sin.[90] We can simplify.

Faith comes down to a disposition of the human heart. We all understand that there is a link between what one really believes in the heart and what one says. Notwithstanding, human beings have the capacity to lie. However, trying to lie to God is astonishingly stupid. In his letter to the church at Rome the Apostle Paul explained salvation as follows:

"If you confess with your mouth the Lord Jesus and believe in your heart that God has raised Him from the dead, you will be saved. For with the heart one believes unto righteousness, and with the mouth confession is made unto salvation." (Romans 10:9-10)

A clear example of the saving power of God's grace occurred on the day Jesus was crucified. As He hung on that bloody cross between two thieves, one ridiculed Him while the other sought mercy.[91] The one who sought mercy said, *"Lord, remember me when You come into Your kingdom."* (Luke 23:42) This fellow wasn't a theologian or scholar. He had no credentials. He knew nothing about prevenient, justifying or sanctifying grace. He would get no opportunity for baptism or other religious activities. He just believed in his heart that Jesus was who He claimed to be. Jesus told this man, *"Assuredly, I say to you, today you will be with Me in Paradise."* (Luke 23:43) This thief did "land safe on that happy shore."

The episode described above can be compared with so-called "death-bed" conversions. Last minute confessions result in the confessor landing safe on that same happy shore alongside those with decades of Christian service. Sadly, some of those who have born the heat of the day resent this expression of God's grace.[92] Perhaps their resentment springs from lingering notions of rewards based on merit discussed earlier. After all, shouldn't those who work harder or longer be rewarded more than those who do less? Grace, grace, grace — not works. It doesn't matter whether others approve or not. God is the author of salvation. And God's grace is truly amazing. There's always hope, right up to the very last breath. But when the door of salvation closes, it closes forever.

The thief on the cross didn't have *any* good works. But he's got a pretty good zip code today.

Epilogue

Gauguin asked, "Where do we come from?" Science claims and evidence confirms that the atoms which form the human body came out the Big Bang by way of the stars.

He asked, "What are we?" Science claims and evidence confirms that we are bipedal primates with mixed ancestry. We are descendants of anatomically modern Homo Sapiens who evolved in Africa and immigrated into the Middle East where they mixed with Neanderthals. Scripture claims we are the descendants of Adam and Eve. I have argued we are both. Answers to his first two questions are universal, which means they apply to everyone.

The artist asked, "Where are we going?" The answer to his third question comes in two parts. The first part is that upon death, the atoms of the body will return to the dust from whence they came. This is the universal part. The second part of the answer is not universal, which means the answer varies from person to person. Specifically, some souls will "land safe on that happy shore," but others will land in hell.

Gauguin posed his three questions in the first person, plural pronoun "we." Because the second part of the answer to his third question is not universal, perhaps we should recast his last question in the first person, singular pronoun.

Where am I going?

Notes

[1] Matthew 16:26 (NKJV)

[2] Debra Silverman, *Van Gogh and Gauguin, The Search for Sacred Art*, (New York: Farrar, Straus and Giroux, 2000) pg. 121.

[3] Ibid_____ pg. 373.

[4] Alan H. Guth, *The Inflationary Universe*, (Cambridge: Perseus Books, 1977) pg. 2.

[5] Lawrence M. Krauss, *A Universe From Nothing: Why There is Something Rather Than Nothing,* (New York, Free Press, 2012) pg. 156.

[6] Hermann Gunkel, W. H. Carrots, trans. *Legends of Genesis*. (Chicago, The Open Court Publishing Company, 1900) pg. 159.

[7] Stephen Weinberg, *The First Three Minutes: A Modern View of the Origin of the Universe,* (New York: Basic Books, 1977) pg. 131.

[8] Eric R. Kandel, *In Search of Memory: The Emergence of a New Science of Mind* (New York, W. W. Norton & Company, 2006) pg. 68.

[9] Francis Crick, *The Astonishing Hypothesis: The Scientific Search for the Soul,* (New York, Simon and Schuster, 1995) pg. 257.

[10] Richard P. Feynman, *The Meaning of It All: Thoughts of a Citizen Scientist* (Addison Wesley, 1998) pg. 106.

[11] Martin Rees. *Our Cosmic Habitat*, (Princeton, NJ, Princeton University Press, 2001) pg. 75.

12 Ibid_____ pg. 124.

13 H.A. Harper, V.W. Rodwell, P.A. Mayes, *Review of Physiological Chemistry, 16th ed.* (Lange Medical Publications, Los Altos, 1977).

14 R. Sender, S. Fuchs, R. Milo (2016) *Revised Estimates for the Number of Human and Bacteria Cells in the Body.* PLoS Biol 14(8): e1002533.

15 Govert Schilling, *Flash! the Hunt for the Biggest Explosions in the Universe,* (Cambridge, MA, Cambridge Univ., 2002) pg. 262.

16 Martin Rees, J*ust Six Numbers: The Deep Forces That Shape The Universe,* (New York, NY, Perseus Books, 2000).

17 Ian Tattersall, *The Strange Case of the Rickety Cossack,* (New York: Palgrave MacMillan, 2015) pg. 215-216.

18 Israel Hershkovitz, et.al. *A Middle Pleistocene Homo from Nesher Ramla, Israel,* (Science, 2021) science.abh3169.

19 Erwin Schrodinger, *What Is Life?* (Cambridge: Cambridge University Press, 1944).

20 R. Sender, S. Fuchs, R. Milo (2016) *Revised Estimates for the Number of Human and Bacteria Cells in the Body.* PLoS Biol 14(8): e1002533.

21 Jack Challoner, *The Cell: A Visual Tour of The Building Block of Life,* (East Susex, Ivy Press: 2015) pg. 166.

22 D. J. G. Mackay, I. K. Temple, *Human imprinting disorders: Principles, practice, problems and progress.* Eur J Med Genet. 2017 Nov;60(11):618-626. doi. Epub 2017 Aug 14. PMID: 28818477.

[23] Y. Le Bouc, S. Rossignol, S. Azzi, V. Steunou, I. Netchine, C. Gicquel, *Epigenetics, genomic imprinting and assisted reproductive technology (*Ann Endocrinol, Paris). 2010 May;71(3):237-8. Epub 2010 Apr 2. PMID: 20362968.

[24] Hans Eiberg, et. al., *Blue eye color in humans may be caused by a perfectly associated founder mutation in a regulatory element located within the HERC2 gene inhibiting OCA2 expression.* (Human genetics, 2008; 123 (2): 177).

[25] Ian Tattersall, *Masters of the Planet* (NY, Palgrave Macmillan, 2012) pg. 199.

[26] Genesis 3:19

[27] James 1:5

[28] Abraham Pais, *'Subtle is the Lord...': The Science and the Life of Albert Einstein*, (Oxford, Oxford University Press: 1982).

[29] I Peter 3:15

[30] I Corinthians 14:33

[31] Genesis 2:7 (KJV) Other translations report that Adam became a living "being".

[32] Richard Feynman. *The Meaning of It All: Thoughts of a Citizen-Scientist,* (Reading, MA. Addison-Wesley Books: 1998) pg. 3.

[33] Matthew 8:4, Luke 16:31 and Luke 24:2 for examples.

[34] Robert C. Dentan, *The Story of the New Revised Standard Version,* (Princeton, Princeton Seminary Journals, The Princeton Seminary Bulletin: 1990) pg. 220-221.

[35] http://www.lostcity.washington.edu. accessed April 2, 2016.

36 Kelley, D.S., J.A. Karson, S.P Sylvia, et. al.. A Serpentinite-hosted Submarine System: The Lost City Hydrothermal Field. Science 307, 1428-1434 (2005).

37 Carl R. Woeses and George E. Fox. Phylogenetic structure of the prokaryotic domain: The primary kingdoms. PNAS. vol. 74 no. 11, Nov. 1, 1977.

38 The Abridged Brown-Driver-Briggs Hebrew-English Lexicon of the Old Testament: from A Hebrew and English Lexicon of the Old Testament by Francis Brown, S.R. Driver and Charles Briggs, based on the lexicon of Wilhelm Gesenius. Boston; New York: Houghton, Mifflin and Company.

39 Matthew 10:28

40 Matthew 7:13

41 Matthew 10:28; 7:13 and John 3:16

42 Matthew 25:41

43 Ezekiel 18:4

44 Phys Life Rev. 2014 Mar;11(1):39-78. doi: 10.1016/j.plrev.2013.08.002. Epub 2013 Aug 20. PMID: 24070914.

45 Romans 3:23

46 II Peter 3:9

47 John 10:27

48 Exodus 3:1-4:17

49 Israel Hershkovitz, et al, *Nature*, http://dx.doi.org/10.1038/nature14134;2015.

50 Genesis 5:1-5

[51] Maimonides scholar Shlomo Pines said, "Maimonides is the most influential Jewish thinker of the Middle Ages, and quite possibly of all time" (*Time* magazine, December 23, 1985).

[52] www.chabad.org/library/article_cdo/aid/889836/jewish/Maimonides.h.

[53] Moses Maimonides, *The Guide for the Perplexed*, translated by M. Friedlander [1903], Part III, Chapter 8.

[54] Genesis 4:1, 4:17 and 4:25

[55] Jennett, Karen Diane. *Female Figurines of the Upper Paleolithic, Honors Thesis Presented to the Honors Committee of Texas State University-San Marcos*, San Marcos, Texas. May 2008:13-15.

[56] Isaiah 55:11

[57] Romans 5:12

[58] II Peter 3:8

[59] Genesis 3:21 See entry # 2290 in Strong's Enhanced Lexicon, and כְּתֹנֶת in the Abridged Brown-Driver-Briggs Hebrew-English Lexicon of the Old Testament.

[60] Ephesians 2:1-10

[61] John 1:29

[62] Revelation 13:8

[63] Exodus 32:1-6

[64] Ezekiel 18:20

[65] Romans 3:23

[66] Hebrews 9:22

67 Jeff A. Brenner. http://www.ancient-hebrew.org/articles_cainandabel.html. Accessed April 11, 2019.

68 Partial Abstract published in *Laboratory Medicine:* "A case of disputed paternity is presented in which there is evidence that dizygotic twins were actually half siblings sired by different men. Case reports such as this one are rare, but superfecundation may actually be common in man. The terminology, history, and biology of human superfecundation and related phenomena are discussed." Laboratory Medicine, Volume 17, Issue 9, 1 September 1986, Pages 526-528.

69 https://medicalxpress.com/news/2020-10-early-trauma-metabolism.html Accessed December 10, 2020.

70 1 John 3:11–12

71 Genesis, chapter three

72 Genesis 6:6

73 2 Samuel, chapter 11

74 1 Jn 3:11–12

75 Exodus 20:14

76 Leviticus 20:10

77 Luke 3:38

78 Deuteronomy 2:10-11; 3:11; Joshua 12:4. In addition, the Psalmist recounts the defeat of Og in Psalm 135.10-11.

79 Flavius Josephus, *The Complete Works of Josephus*. Translated by Wm. Whiston/Foreword by Wm. S. La Sor, (Kregel Publications:1981) pg. 83, 165.

[80] Isabelle Mansuy, University of Zurich, August 31, 2020. https://neurosciencenews.com/epigeneitc-trauma-metabolism-17177/ Accessed October 20, 2020.

[81] I Timothy 6:10

[82] Genesis 6:5-6

[83] Abridged Brown-Driver-Briggs Hebrew-English Lexicon of the Old Testament. 306.2, 307.102.

[84] Enhanced Strong's Lexicon, entry # 266.

[85] John 6:44

[86] Exodus 20:8

[87] Enhanced Strong's Lexicon, entry # 3340.

[88] The Theological Dictionary of the New Testament, Abridged in One Volume (Little Kittle) is considered by many to be the best New Testament Dictionary ever compiled. TDNT provides extensive treatment of the Pist Group of Greek words.

[89] Galatians 2:1-21

[90] I Corinthians 5:1-13 describes sin going on in the early church. Modern denominations continue to struggle with some of these issues — especially sexual sins.

[91] Matthew and Mark each report that both thieves mocked Jesus. Matthew 27:44, Mark 15:32b

[92] Matthew 20:1-16

www.ingramcontent.com/pod-product-compliance
Lightning Source LLC
Chambersburg PA
CBHW072151070526
44585CB00015B/1084